PROJECT PLANNING & CONTROL FOR CONSTRUCTION

David R. Pierce, Jr.

PROJECT PLANNING & CONTROL FOR CONSTRUCTION

David R. Pierce, Jr.

R.S. MEANS COMPANY, INC.
CONSTRUCTION CONSULTANTS & PUBLISHERS
100 Construction Plaza
P.O. Box 800
Kingston, Ma 02364-0800
(617) 747-1270

© 1988

Printed in the United States of America

10   9   8   7   6   5   4   3   2   1

Library of Congress Cataloging in Publication Data

ISBN 0-87629-099-3

Dedication

This book is dedicated to the author's colleagues at the School of Building Construction, University of Florida; the Department of Industrial Construction Management, Colorado State University; and the Department of Construction Management, California Polytechnic State University, who have worked so hard over the years to create the profession of Construction Management.

It is also dedicated with love to Jennifer and David.

Table of Contents

Foreword   xi

Acknowledgments   xiii

Introduction   xv

Chapter 1: What Is Project Management?   1
   What Is Management?   3
   What Is Project Management?   4
   Why Use Project Management?   7
   The Benefits   7
   Variables that Affect Project Management   9
   Use of Computers in the Construction
     Industry   10

Chapter 2: The Project Control Cycle   13
   Basic Management Functions   16
   Introduction to the Sample Building Project   18

Chapter 3: Pre-Construction Planning   29
   Identifying Key Personnel   32
   Providing Data to the Project Team   32
   Contract Document Review   32
   Estimate Review/Estimator Meeting   33
   Initial Project Team Meeting   35
   Pre-Planning with Other Parties   36
   Setting up Management Procedures   40

Chapter 4: Introduction to Scheduling   45
   The Critical Path Method   47
   Learning CPM Techniques   48

Chapter 5: Planning the Project   51
   Breaking the Job Down into Activities   53
   Establishing the Sequence of Work   57
   Expanding the Network   58

Chapter 6: Scheduling the Project   67

  Estimating Durations   69

  Adjusting Activity Times   87

  Calculating Overall Job Duration   89

  Advanced Calculations   96

Chapter 7: Management of Submittal Data
and Procurement   105

  The Source of the Problem   107

  Basic Procurement Procedures   108

  Key Elements in Successful Procurement   108

  Record Keeping and Tracking   110

  Scheduling the Procurement Activities   119

  Reporting   120

  Follow-up on the Information   120

Chapter 8: Organizing the Schedule   121

  Why Organize the Schedule?   123

  Levels of Detail   124

  Tasks Required to Provide the Right
    Information   126

  Types of Coding Schemes   126

  Coding Schemes for a Project   127

  Computer Coding   128

  Using the Coding Structures to Get
    Information   132

  Selection   132

  Sorting the Activities   136

  Summary of Activities   136

Chapter 9: Monitoring and Controlling the Project   141

  Communication   143

  Monitoring Progress and Schedule Update   144

  Comparing Progress to Goals   152

  How to Find Out Why the Job Is Behind   155

  Taking Corrective Action   155

Chapter 10: Maintaining the Schedule Records   157

  Initial Development Phase   159

  Inputting Data to the Computer   167

  Running the Trial Calculations   167

  Schedule Documentation   170

  Providing Information for Construction   170

  Reporting and Updating the Schedule   171

Chapter 11: Resource Management   173

  Close Management of Resources   175

  The Resource Management Process   176

  Development of the Resource Profile   176

  The Value of Computerization   180

  Adjusting the Schedule to Improve
    Resource Expenditures   180

  Managing Cash Flow   184

Forecasting and Managing Cash Flow 188
Practical Aspects of Resource Management 191

Chapter 12: Time/Cost Tradeoffs 193
Reasons for Reducing Overall Project Time 195
Relationship of Project Cost to Time 196
How to Reduce Project Time 196
Practical Considerations 201

Chapter 13: Project Cost Control 205
Project Cost Coding System 208
Elements of a Project Coding System 208
Specific Tasks in Project Cost Control 210
Other Cost Control Issues 225

Chapter 14: Summary and Conclusions 229

Appendix A: Sample Project Estimate 235

**Appendix B: Sample Logic Diagrams and Computer
Reports** 241

Appendix C: C.S.I. Masterformat Divisions 267

Foreword

Today's construction projects are more complex than ever before. Too many must pay the price of time and cost overruns resulting from a lack of proper planning, organization, monitoring, and control. The aim of this book is to help project managers and members of the project team avoid these problems by adapting effective, proven methods for planning and controlling construction projects. The book presents a view of the complete project planning and control process — from basic management principles and pre-construction planning through scheduling, monitoring, record keeping, and cost control.

The first two chapters outline the basic tenets of effective project management, and offer many tips for making the overall process more effective. A sample building project is introduced at the end of Chapter 2 and used throughout the succeeding chapters to illustrate various principles of project planning and control. Chapter 3, "Pre-Construction Planning," identifies key project team members and spells out the essential steps for getting a project started on the right footing.

A good schedule is the primary tool of the successful project manager. Chapter 4 is an introduction to Critical Path Method scheduling — the basic techniques, benefits, and pitfalls to avoid. Chapters 5 and 6 present the specific, detailed tasks of planning and scheduling the actual construction of a project, while Chapter 7 is concerned with scheduling administrative tasks for procurement of materials.

Hundreds of tasks must be precisely choreographed if a project is to run smoothly — on time and within budget. Chapters 8 and 10 describe how to efficiently organize and manage the data generated by the scheduling process.

Delays and other complications inevitably occur on construction projects and dealing with them effectively is a key ingredient in successful project management. Chapter 9 addresses the ongoing tasks of monitoring project time, and presents specific actions that can be taken to bring the project back into line.

Chapters 11 and 12 are devoted to advanced topics — how to use the schedule to manage the resource expenditures on the job, and how to make the difficult and complex decisions involved in shortening the duration of the project. Chapter 13 presents an effective cost control system for a project, and explains how to run and use it.

The three appendices include a partial estimate and schedules (CPM logic diagrams and computer reports) for the sample building project. A listing of the Construction Specifications Institute's MASTERFORMAT headings is also provided for reference.

Acknowledgments
The author would like to express grateful appreciation to all of the individuals and companies who contributed to the preparation of this book by so willingly and enthusiastically sharing their experience and knowledge over the course of the author's career in construction teaching and practice.

Particular appreciation is due to Paul Miskel for his cogent comments and editing of the technical material, and to Ted Wetherill for his support during the writing process. A special thanks is given to Mary Greene for turning the author's construction and academic jargon into a clear and readable product.

Introduction

Project planning and control is a three-step process that requires the involvement of all parties on a construction job. It demands that the participants look ahead at the work to be done, plan strategies for getting it done, and then monitor the work to ensure completion according to plan. Each step requires the commitment and dedication of all project team members. All are essential to the successful outcome of a construction project, and to the long term success of a construction firm.

The first step is establishing a plan of action for carrying out the construction work. This involves breaking the job down into a series of manageable sub-parts, deciding on an order of placement, and determining the time and cost for each part and for the job as a whole. This early planning results in a "road map" to be followed by all persons and companies on the job — to finish the project on time and within budget. There is no one best plan for each project, just as there is no one correct estimated cost. There is only a choice among possible plans. The task of the project manager is to see that a reasonable and effective plan is devised, one that reflects the intentions and desires of those who will be responsible for its accomplishment.

After the plan has been developed and officially accepted, it must be effectively communicated to all those on the job. The quality and plausibility of this plan and the degree to which management personnel are involved in its development set the tone for the work itself.

Once the job is under way and the plan is being carried out, events inevitably occur that will force changes in the job. To effectively cope with these events, the project manager must ensure that the second management step is carried out.

The second step in project planning and control is regular monitoring of the job. This includes comparing the progress of the work to the original plan to see which aspects of the job are going according to plan, and which are not. There are two considerations: time and cost. To check time progress, individual parts of the job are tracked to determine if they are completed within the *time limits* of the schedule. For a cost check, the *actual* costs of construction must be determined and compared to the *budgeted* costs for that point in the job. With information gathered from this tracking and monitoring, the project manager must find out where the job is off track.

Finally, in the third step, the project manager must determine the causes of any delays and cost overruns that have occurred, and then follow through. This means working with the personnel responsible for the job's progress to ensure that corrective action is taken to bring the troubled parts of the job back into line. In order to do this, the project manager must involve all the working managers in the problem solving process to determine causes and corrective actions. This process frequently involves re-planning and aggressive follow-up. These actions are essential, however, if the job is to be properly controlled and a late and unprofitable finish avoided.

Project planning and control is a difficult task, but one that must be performed well to realize both short- and long-term success in the construction business. The process requires taking a close look at field procedures and plans, and at the ways in which field managers and subcontractors are managed and scheduled. The rewards of effective project management are time saved and money earned. When vigorously pursued by the project manager, these methods help prevent the problems of late job completions and cost overruns. This book is intended to provide all project management professionals with effective skills and tools to carry out these planning, monitoring, and follow-up tasks, and to show how these tasks can be carried out with the least possible cost and effort to the individuals and their companies.

1

What Is Project Management?

What Is Project Management?

The goal of this book is to help the reader become a more effective construction project manager, or to help a construction company carry out more effective project management and control. To achieve this goal, we must start with an understanding of project management, or project control. These two terms have virtually the same meaning, that is, taking whatever actions are required to plan and build a project on time and within budget, while satisfying the owner's goals. Clearly, a contractor who succeeds in fulfilling these aims will remain in business and is likely to prosper. Now, what are the nuts and bolts of project management?

What Is Management?

- It is the planning, organizing, staffing, directing, and controlling of resources to achieve a company's goals.
- It is deciding for individuals and groups the use of resources and implementing such decisions.
- It is controlling and directing human resources toward a goal.
- It is the rational, systematic process of decision making, to achieve specific goals.

These general definitions introduce some key elements:

Goals
Process
System
Decisions
Rational Process
Personnel
Resources
Planning
Control

At the very beginning, we recognize that **goals** are always involved. In a philosophical sense, without a set of goals, there is no point in even taking actions. In practical terms, a construction company or manager must have profit as an overall goal. Establishing smaller and more focused short-term goals is a very important part of the overall task of project management.

Process can be defined as a set of continuing actions over time. The management process must be carried on continually throughout the life of the job or company. Management must be done in a **systematic,** which means an orderly, regular, and dependable way, using a set of established procedures or methods.

Management also means that **decisions** must be made in order to achieve the goals of the project or company. These decisions must be made **rationally,** that is, based on facts, not hunches or inaccurate information. One of the primary reasons for setting up a systematic, orderly method of management is to deliver accurate, timely information to the decision makers on the job.

Personnel are always a management consideration. On a job site, there are craftsmen, subcontractors, suppliers, designers, and owner's representatives, among many others who must work together in the process of getting a project built. The necessary personal and organizational interactions make the process complex and challenging. The decisions made by managers must take into account the requirements, opinions, and attitudes of all these concerned parties. A good project manager brings these parties together regularly, to encourage a shared effort toward achieving the project goals.

We must also be concerned with the expenditure of **resources** as the job progresses. Labor must be hired, and effectively and efficiently controlled; material bought at the right time and price; subcontractors employed and properly scheduled; and owner-furnished materials supplied and properly installed. Again, the system we establish must provide for the use of these resources in the best possible way toward achieving the project goals.

Once a sound project management system is in place, the project manager may use it to carry out the more specific functions of **planning** and **controlling.** *Planning* can be defined as scheduling the tasks that will accomplish the goals of the project. This means establishing realistic schedules and budgets, coordinating resources to get the work done, and — most importantly — making sure everyone knows what the plan of action is.

Controlling is the final action in the management process. To achieve and maintain control, the project manager monitors the progress of the job. When short-term goals are not being met, the project manager must take action to get everything back on track.

What Is Project Management?

The previously listed concepts are the basis for effective construction project management. The tasks listed below grow out of these concepts and are the focus of this book.

1. Establish and focus on goals, that will be general at first, then increasingly specific and job-oriented.
2. Establish an effective management process, that will operate in a systematic manner.
3. Use this management process, or system, to make the best possible rational decisions for efficiently using resources, coordinating personnel on the job, and planning and controlling the work throughout the life of the job.

Having defined the three major project management tasks, we can begin to focus on the more specific jobs of the project manager.

1. Setting Goals

The first task of the construction project manager is establishing goals. Many goals have already been set by the project estimate and contract documents. A primary purpose of the estimate is to arrive at a cost for the project, while the contract establishes the time required for completion. Neither of these goals — cost or time — can be altered in any significant way by the project manager. However, the project manager can set intermediate goals for the construction process, goals that meet the ultimate requirements of cost and time.

2. Creating a Project Management System

After setting goals for the project, the next task of the project manager is to establish control through an effective management system. There are two approaches to project management. They are: 1) proactive — aggressive management ensuring that the job proceeds as planned by the manager; or 2) reactive — spending money and reacting as events occur, letting circumstances run the job. The latter, reactive scenario tends to occur if the manager fails to set up a properly organized, thorough, and methodical management system. This approach almost certainly guarantees that a job will overrun time and budget allowances, and generally cause much grief to all concerned.

The management system should be designed to address the following elements:

Time: A plan of action must be established to assure the work is done in the correct order or sequence and within the time allowed.

Cost: The work must be performed efficiently if the contractor's goals are to be met.

Resources: It must be determined in advance when and how much of each resource (such as particular categories of labor, equipment, or materials) is needed to do the work. One must then ensure that the resource is provided when and where needed. Resource management supports the effort to control time and cost. The information used by the project manager to perform this task is developed from the time and cost information data.

Finances: Ultimately, time and resources translate into dollars. Thus, the financial control function means accurately predicting the amount of cash needed to support all the work done on the job.

3. Managing the Project

The third task of the project manager is to manage the project as it proceeds, using the project control system to best advantage. In the simplest terms, is the project heading in the right direction to meet its goals? Project control is best illustrated as a *feedback loop,* shown in Figure 1.1.

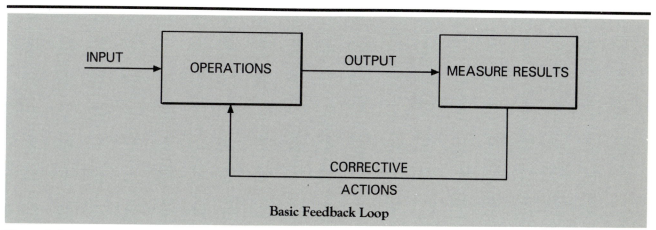

Basic Feedback Loop

Figure 1.1

Managing the project begins with the input of data on labor, materials, and equipment — the resources used to build the project. As the work is performed, the output, or productivity, is measured to see if it is meeting the goals set for the project. If it is not, corrective action must be taken.

Project control can be illustrated by comparing it to a feedback loop used in machine control. For example, a cruise control on a car measures the speed of the car, compares it to the preset speed, and if there is a variation, either adds or reduces throttle to return the speed of the car to the preset limit. If the cruise control fails to function properly, the car either takes a long time to reach its destination, or the driver gets a ticket. Using feedback to manage and control a construction project, however, requires a few more steps. The project manager must perform the following tasks.

Plan: Realistic, usable schedules and budgets must be established for all phases of the job. These guidelines will serve as a "blueprint" for building the job. The schedules and budgets should be based on the original estimate and contract requirements. They must reflect the commitment of the people who will have to carry them out.

Communicate: Once developed, the plans must be communicated clearly and effectively to the people who will be executing them. Emphasis must be placed on providing clear, usable visual displays, particularly for scheduling.

It is also wise to recognize that the professionalism shown by the project manager in planning and communicating on the job site has a very real effect on employee morale and effectiveness. A sloppy plan, poorly organized and executed, gives employees the impression that they work for a slipshod organization. Pride in their work will be affected accordingly.

Monitor and Control: After the plans have been developed and communicated, they must be carried out by project personnel. Realistically, some unexpected events could interfere with the original plan. If this occurs, the project manager must take steps to ensure that the project goals are met. This means taking action to bring the job back in line with the original plan, or revising the plan to fit the new situation.

If the project manager is to effectively deal with delays, the management system must provide him with the most current information. This monitoring function involves collecting data on time and cost, and comparing this information to original projections.

Once the project manager is aware of the current job status as compared to the original plan, actions can be taken to meet the original goals. These actions can range from adding more crews to speed up sheetrock installation, to completely changing the installation sequence of complex formwork.

Two other points should be made about the feedback loop. First, construction job sites can be very busy places, with many activities going on at one time. Therefore, it is important that the management system be *exception-oriented*. That is to say, the system should be designed to specifically point out those items that are at variance with the plan, and to essentially ignore those that are proceeding on schedule. Without an exception-oriented system, the project manager is in danger of being overwhelmed with detail, while key areas may be overlooked.

Second, the information provided must be *timely,* so that problems are caught and recognized early in the game. Problems on a job tend to worsen at an accelerating rate. It is important to catch them before they have a chance to become major disasters.

The key to monitoring and control by project management is making frequent checks of job status and, if necessary, taking action to ensure that the project's goals are met. Failing in either checking or acting will result in a failure to meet the project goals.

Why Use Project Management?

In the previous sections, we have recommended an overall strategy for project management and control. There is still, however, some feeling in the construction industry that all this management and paperwork is not really necessary — that one can monitor the job effectively enough by walking around the site to ensure on-time completion and a profit. In reality, the industry is now such that more effective techniques for control and management on the job are essential.

The recent history of some construction projects is sadly one of the reasons why old methods of project control are no longer effective. Today's construction projects are complex and very different than in the past. For example, building environmental control systems have replaced simple heating systems. Structures may now consist of high-, early-strength, post-tensioned concrete floor systems with shear walls, where we once had simple flat slabs and columns. Windows have become complete "exterior enclosure systems." This means a greater variety of jobs to be done, and a greater percentage of subcontractors. Architects and owners have a much wider range of materials and systems to choose from; thus, very few projects are the same.

The increased variety of construction materials and methods has generated more detail than can be managed effectively by one person. In the past, one individual could carry out most of the management tasks since only the basic trades were involved, there was an architect and an owner, and the contract was straightforward. Today's project manager must coordinate specialized subcontractors and work with design specialists who are, in effect, subcontractors of the architect. Many regulatory agencies have also entered the picture. The total management work load has increased to the point that a team and a comprehensive, well-designed system of control are essential. Contractors who attempt to deal with this new situation in the same old ways have encountered enormous problems.

At the same time, a kind of revolt has been taking place on the part of owners, who are themselves faced with some difficult circumstances. The highly volatile marketplace of recent years, which has been aggravated by periods of high inflation and high interest rates, has put extreme pressure on owners for on-time and in-budget project completion. As a result, owners tend to increase pressure on contractors for better and quicker performance. A contractor not using up-to-date methods is at a serious competitive disadvantage. In some cases, owners have even dictated changes in operating methods to contractors. The demands of clients provide yet one more argument for good project controls.

Good project controls can help keep us out of the legal arena — first by making it more likely that we will perform better, thus reducing the reason for legal action — and second, by providing us with a better set of documents with which disputes may be more easily resolved.

The Benefits

Having reviewed today's situation and having seen the need for an improvement in management, the question arises — why use scheduling and cost control? Above all, they should be used because they have proven

7

effective on construction jobs. In particular, the use of Critical Path Method (CPM) scheduling and cost control systems (and the implementation of these methods using computers) are clearly both workable and cost-effective. This is especially true since the development of microcomputers and software systems which are inexpensive and simple enough to use in a construction trailer.

While a computer and good control systems are not a panacea and will not do the manager's job, they are very helpful in providing a way to set up target plans, track events on the site, and examine alternate ways to correct the schedule of work in the event of deviation from the original plan. A good control system is especially helpful to the manager in pinpointing the problem areas, and therefore helps to manage and reduce the information overload. While such an approach may not guarantee success in terms of cost and time, the intelligent use of control systems definitely increases the odds that success will follow the project manager's efforts.

Better Organization

One of the benefits of using good management systems is that they encourage, or even force, better organization and planning. This is a vital influence since one of the biggest failures of managers in industry is a lack of planning. Effective monitoring and control must start with a workable plan. It has even been said that 75% of the value of creating a CPM schedule is the initial planning that must accompany this process. This statement may be an exaggeration and reflects the difficulty of monitoring with early software systems. Nevertheless, a well organized initial plan starts any job off right.

Using control systems also forces the manager to look at how all of the available resources will be assembled and used. It encourages better purchasing and timing, and reduces wasted motion.

A Good Basis for Coordination

A major problem for many contractors is subcontractor scheduling. A big part of the solution is communication. If the project manager and superintendent maintain an up-to-date schedule and require the subs to attend regular job schedule meetings, all the parties on the job will be operating from a base of common agreement. Also, the regularly scheduled meetings encourage subcontractor participation in the scheduling process. This participation promotes a sense of commitment to the project.

In general, better coordination benefits all of the parties to a project — the owner, contractor, designer, and subcontractor. Delays are prevented rather than reacted to, costs are contained early, claims prevented or resolved earlier and more amicably. The result is a better profit for all.

Management by Exception

As noted, a major challenge for today's project manager is tracking vast amounts of detail. Computers provide an advantage as they are very good clerical and record keeping devices. Most of the computer-based management systems used today take advantage of this capability, and promote the exception-based management approach. The systems can be set up to track all work, but report only those elements of job progress that deviate significantly from the original plan. This leaves the project manager free to devote his construction knowledge and skill to the problem areas, leaving the on-schedule areas to proceed to completion.

Another key point in a computer-based management system is providing early detection of problem areas, thus helping to prevent unpleasant surprises. This early detection is critical to correction, since as we all know, problems on construction sites never get better without attention; they only get worse.

Better Decision-Making

A good, up-to-date management system and the associated techniques will in all cases provide the basis for better decisions on the job. Accurate information is an absolute necessity for sound decisions. Such a system should be designed to display the essential data, and to weigh the effects of alternate plans of action. Many experts suggest that a better term for *project control systems* is *management information systems.* Regardless of the name applied, a good system properly used will result in better decisions, and thus, better results.

Variables That Affect Project Management

While the benefits of project management are clear, it must be noted that a good project management system — whether manual or computerized — cannot be implemented without an investment of money and time on the part of corporate management. It is, however, an investment that brings a definite and positive return. It is certainly worthwhile to review some of the common problems associated with developing better project management. Forewarned, one can at least minimize these problems. Later chapters of this book will cover the procedures and possible pitfalls in more detail.

Personnel

In any change, people must be the first consideration. First, the installation or development of new project management techniques must directly involve the people who will be responsible for the results. Probably the worst possible approach is to simply choose a technique or system, and mandate, "you *will* use this system." The people involved should be recognized as knowledgeable, competent, and concerned about the performance of their job. Their professionalism will also be helpful in choosing and operating any new and better procedure. They must be brought into the decision-making and changeover process.

Job-site personnel may be somewhat intimidated by new methods. For example, a superintendent may fear that a new scheduling or cost system will have the home office looking over his shoulder. Or a project engineer may feel uncomfortable with a new, unfamiliar system, fearing failure due to lack of knowledge. The solution to these kinds of problems lies in: 1) honest dealings with the persons involved, with an emphasis on team improvement and the removal of threatening elements, and 2) training, which will clearly demonstrate the company's commitment to improvement and willingness to continue investing in its employees.

Cost and Organizational Concerns

The project manager must recognize upfront that implementation of better management techniques will cost real dollars and will require some organizational changes and adjustment. As previously noted, the benefits justify the investment of time and money.

Organizational changes clearly involve people. To begin with, any improvement in the system must start with the wholehearted commitment and backing of the company's management. Without it, it is difficult for a single project manager to undertake significant improvement in techniques.

It is also important that company management approach the problem professionally. If, for example, the company president's attitude toward installing and developing a new cost control system seems sloppy and half-hearted, company personnel will perceive that the president does not really care about good cost control. The development of a better cost control system is probably doomed from the start in this circumstance, since the people who have to carry it out will not devote anything like their best effort to a project that they feel the president will not reward.

Another organizational concern is inflexibility, or rather the fear of it, among construction people. The personalities who do well in the construction field are traditionally self reliant and individually competent. They prefer to work with little supervision and to be judged on results, not methods. Also, most field supervisors have more than a few good ideas themselves, which they are willing to share with upper management. A new technique or method should therefore be flexible enough to allow for this kind of individual approach in the field. Such flexibility is relatively easy to achieve with scheduling, but cost control has other restrictions and requirements. Accounting systems must, after all, satisfy company-wide concerns. The emphasis should be on making the burden as light as possible for the field, with the information as accessible and usable as possible.

One final note concerns the importance of communications. When new procedures are being bandied about and changes are in the wind, rumors are inevitable and may hurt morale and cripple the effort to change for the better. Such rumors are best countered by full and open disclosure to project personnel whose lives and professions are directly affected.

Use of Computers in the Construction Industry

It is a fact of life in the construction industry that computers are here to stay. Many contractors have successfully adopted computers for appropriate uses in their organizations and have become more competitive as a result. As is always the case with new management procedures, the implementation and use of computers necessarily involves an investment of both money and managerial time.

The actual investment takes several forms. At the very beginning, management time must be devoted to deciding what kind of hardware and software to purchase and implement. Once these decisions are made, machines and software must be purchased and personnel trained. The training period is apt to be a time of relative inefficiency, since old methods and systems will probably have to be maintained at the same time that the new one is being brought on line. Further, the people being trained in these new methods will be less efficient due to their individual learning curves. After the new procedure is in use, there are ongoing operating expenses.

There is no doubt that computers and software, properly used, are helpful and productive management tools. The problem is how to implement them successfully. Much of this book addresses specific computer applications for project management, but some general points are worth noting here.

First, many companies have never made the move to computerize because they are always waiting for the latest technology or for computer hardware prices to drop a little lower. This approach cannot be easily justified for the following reasons: Generally, computers that are not the latest technology are quite adequate, though they may operate a bit more slowly. They still provide the same benefits to the job. Besides, buying technology that is not brand new means that someone else has already gone through the process and worked out the bugs. In fact, there is a saying in the computer industry: "Never buy the first version of anything."

Second, there may be resistance to the use of computers by company personnel. This is a natural, but avoidable phenomenon. Most importantly, the transition must be carried out in a non-threatening way. The greatest problems are connected with fear of a loss of jobs, either through redundancy or not being able to adapt—to the new systems or to computers in general. The fact is, the introduction of computers very seldom results in a reduction in the number of employees. Instead, they allow the same employees to do the same work (or more) more efficiently. These facts must be communicated clearly and openly to the employees who must deal with the new system. Ample training must also be provided to overcome the fear of not being able to work within the computer system.

The users of a new computer system should be included in the selection and design process. The system should be introduced in a gradual, methodical way to ensure minimum disruption of ongoing work. It is especially important to maintain the old systems until it is certain that the new one works.

Third, all procedures should be properly working by hand before they can be computerized. For example, record keeping for change orders cannot very well be input to a computer when adequate hand records are not being kept. However, computerization may serve as the "push" the project team needs to get a set of procedures in order. If by automating, the team is forced to take a long, hard look at how change orders are handled, then the computerization effort will have a twofold benefit.

It is probably best to approach computerization cautiously but not fearfully, and to take the attitude that slower is better. In general, it is best to begin with simple, straightforward applications, such as word processing and spread sheets, working up to more complex applications like scheduling, and ultimately perhaps to data base record keeping for submittal data and transmittals.

2

The Project
Control Cycle

2

The Project Control Cycle

We have discussed monitoring and controlling, functions that are at the heart of the project manager's job. The goals of monitoring and controlling can also be used as a model for setting up the operations of our project management system. The simple feedback loop (see Figure 1.1) does not, however, cover all of the details that must be dealt with on an actual project. The *Project Control Cycle,* shown in Figure 2.1 illustrates the practical workings of project management.

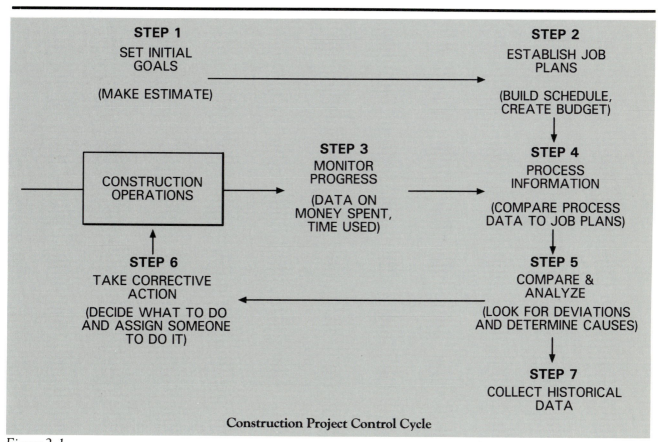

Construction Project Control Cycle

Figure 2.1

Basic Management Functions

It should be noted that this book covers time (scheduling), and cost (job cost control) separately. While integrated estimating, cost, and scheduling systems are available, most contractors perform the two functions at the same time using separate systems. Many firms use computerized cost control, but manual scheduling. Scheduling may be done on the job site, while cost control is maintained out of the home office. Any method is valid so long as the basic functions are performed effectively.

Step 1 — Set Initial Goals

The first step in the project control cycle is setting initial goals. This step typically occurs before the job is even awarded to the contractor. The initial goal is generally no more than a profit goal for the project, in the form of an estimate. Regardless of whether that estimate is conceptual or fixed, or something in between, it serves as a limiting factor, along with the time allowed in the contract documents. Simply put, no future budget for construction should exceed the costs anticipated in the estimate, nor can the time planned in a schedule exceed the number of days permitted under the terms of a contract. The detailed development of estimates is not within the scope of this book. However, using the estimate and contract documents to develop the intermediate goals and job plans is covered in detail.

Step 2 — Establish Job Plans

This second step — establishing job plans — typically occurs after the job has gotten underway, and crews and equipment have moved onto the site. It may not seem advisable to start without a control mechanism completely in place. However, initial conferences will typically have been held during the period right after contract award. Information gleaned during this time may provide a good basis for initial decisions. Such decisions are later developed into detailed job plans.

Establishing job plans for scheduling is done in a three-step process. First, the overall job is broken down into workable parts or *activities*. These activities can be analyzed and planned independently for maximum efficiency. Second, the activities are strung together in a realistic order of work, which is then converted into a logic diagram, or network. The third and final step involves network calculations to determine at what time and on what dates each activity should occur. The final result is a comprehensive plan which serves the following two functions: it is a guide to action by all those involved with building the job, and it can later be used to effectively cope with the inevitable changes that will occur.

The detailed cost plans exist in the form of a budget developed from the cost estimate. This budget breaks the job down into workable parts. The format of the budget also differs from that of the estimate.

Step 3 — Monitor Progress

After the job is underway and detailed plans have been drawn up in the form of a budget and schedule, the job monitoring system must be established and used by the project managers. The first part of this process is carried out on the construction site at regular intervals, and involves monitoring the actual events which occur on the project, to be compared with the schedule later in the cycle.

Schedule monitoring is done on an activity-by-activity basis, again reflecting the very important concept of dealing with workable-sized units of the job at all stages. Typically, each activity is labelled with the following information: start time, duration of work, and anticipated final completion.

In terms of cost, data is recorded as to the actual work done and the expenditures required to get the work done. These expenditures are always broken down into detail consistent with the budget breakdown. This information is typically subdivided further into labor, equipment, material, and subcontract costs for further analysis and control.

Step 4 — Process Information

This activity occurs throughout the monitoring and control process. A computer or manual procedure is used to manipulate the data collected during the monitoring phase. The data is set up so that it can be compared with the plans developed earlier. This processed information enables the project manager to determine whether or not the project is deviating from the planned order or rate of progress, and if that deviation is significant enough to warrant action.

The key element in this information processing phase is the management of the project control *system,* as distinct from managing the project. The processing and use of the plans and monitoring of data depends in large part on having logical and workable coding systems, both for scheduling and cost reporting. Also needed is a regular, efficient, and workable procedure for quickly and accurately developing comparison reports for management.

As projects grow larger, managing the project control system becomes more and more a full time job for a specialist. However, this does not mean that a project management system cannot be operated without specialized job site personnel. The key is having a system that is appropriate for the job at hand.

Step 5 — Compare and Analyze

In Step 5, the project managers review the information developed by the system in the last stage in order to determine the actual state of the project. Report formats must be selected for typical job decisions, and other methods established for efficient exception reporting. The greater the efficiency, the more the project manager's efforts can be directed to those areas of the project which are most in need of management attention.

Step 6 — Take Corrective Action

Taking corrective action represents the final step — acting to rectify an aspect of the job which is not going according to plan. A complete evaluation of Step 6 would include a wide range of topics, since the project manager must deal with technical questions relative to the actual delay or cost overrun.

It is fairly common in construction for deviations from plan to be noted, but not followed up by project personnel. This is equivalent to not monitoring the job at all, and means that not only is the effort spent in developing a project management system wasted, but the job is also likely to end up behind schedule and over cost. The project management feedback cycle must have all its parts working in order for the job to be properly run.

Step 7 — Collect Historical Data

Step 7 occurs on the expanded project control cycle, but does not occur on the general feedback cycle. It involves collecting data on what has happened on the job. Ideally, the results of the job are recorded for two purposes: first, to serve as a basis for estimating future jobs; and second, to serve as a thorough record of actual events in case claims arise. The first function is primarily one of cost, the second one of scheduling.

Introduction to the Sample Building Project

A sample hotel building project is used throughout this book to illustrate various principles of scheduling and project management. The project is a multi-story building with a basement parking garage. The structure is to be built on a confined urban site with limited storage and access. The structure for the parking garage and main floor levels is concrete waffle, and for the upper stories, concrete shear walls, columns, and flat slabs. There is also a main floor wing of structural steel with a steel joist and cellular deck roof. The main floor enclosure system is masonry, while the tower is sheathed in aluminum curtain wall. The interiors of both the main public level and the hotel room tower are commercial finishes typical of hotels throughout the country. The plans for the building are shown in Figures 2.2 through 2.10, and are for illustrative purposes only.

In reality, such a building would require many more sheets of plans, though construction schedules are often developed from schematic or design development drawings of a similar level of detail as these sample drawings. A set of specifications would also be included with an actual plan. However, for the purposes of illustrating scheduling principles, the drawings provide sufficient information. Assumptions have been made for items not shown in the sample drawings that would normally be included in the plans and specifications for a building of this type.

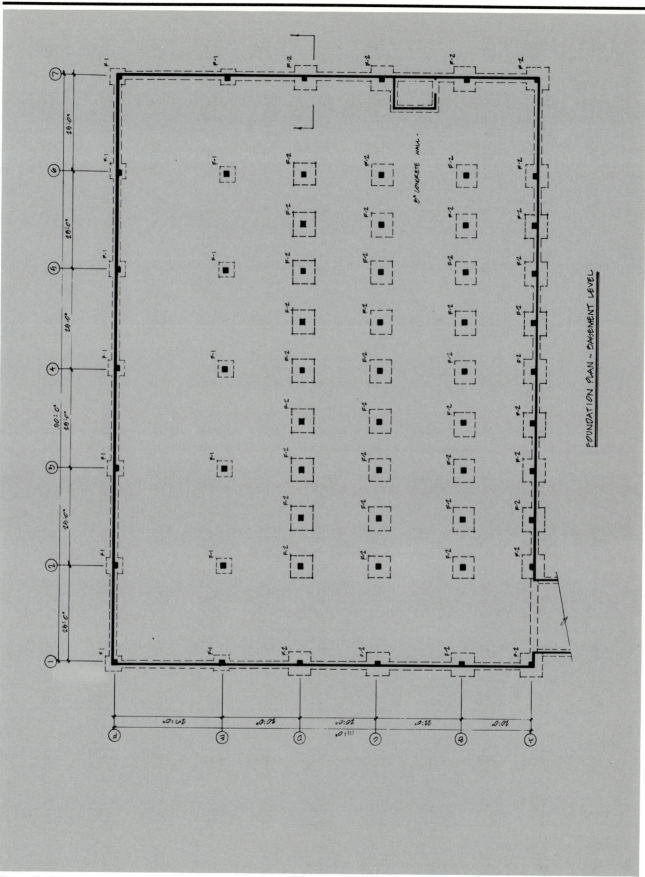

FOUNDATION PLAN ~ BASEMENT LEVEL

Figure 2.2

19

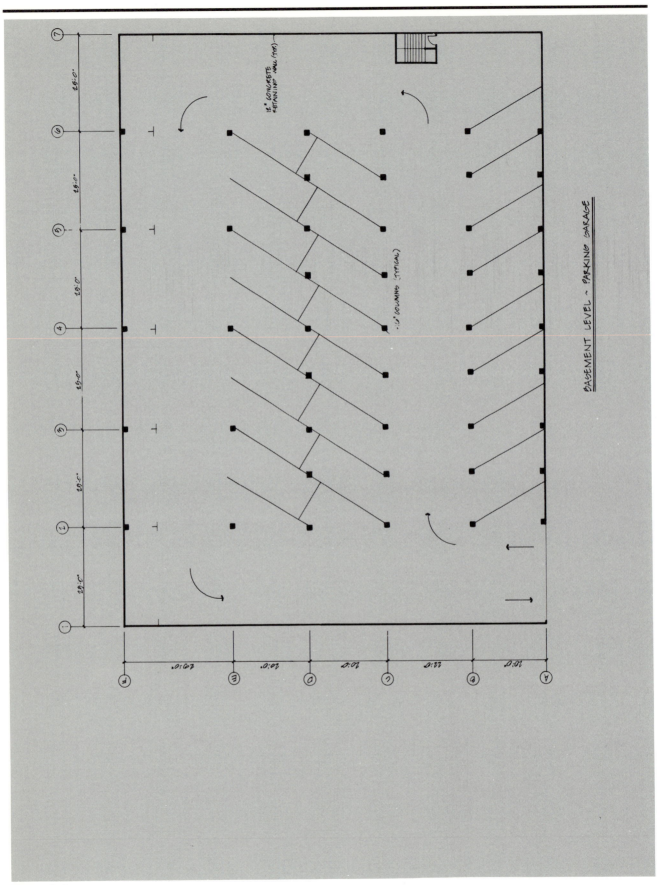

Figure 2.3

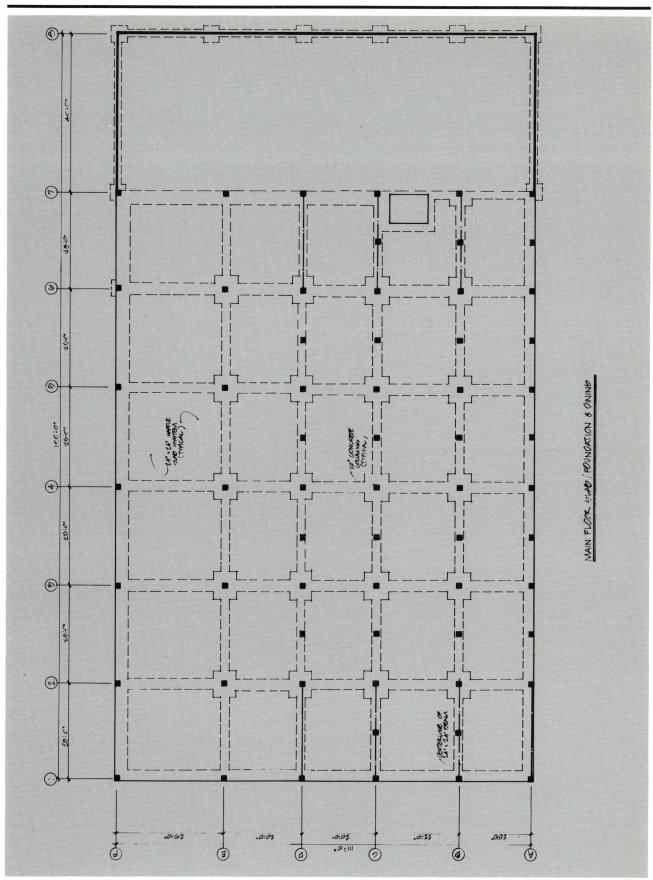

Figure 2.4

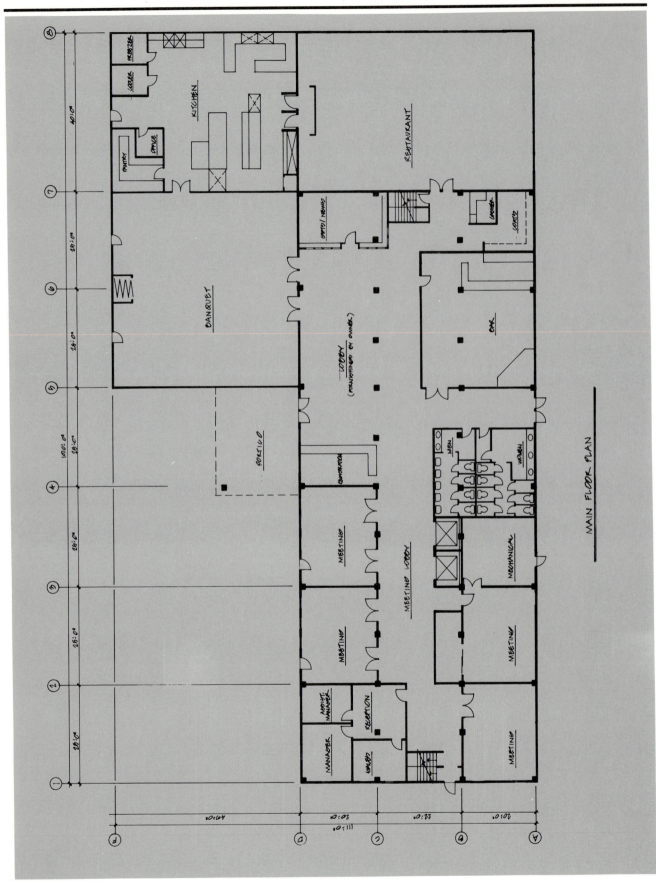

Figure 2.5

22

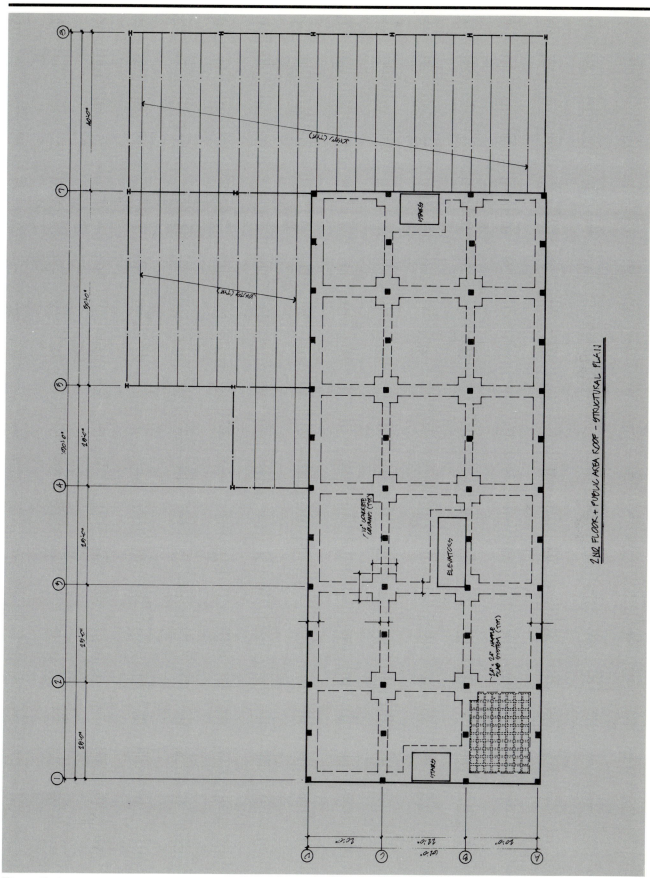

Figure 2.6

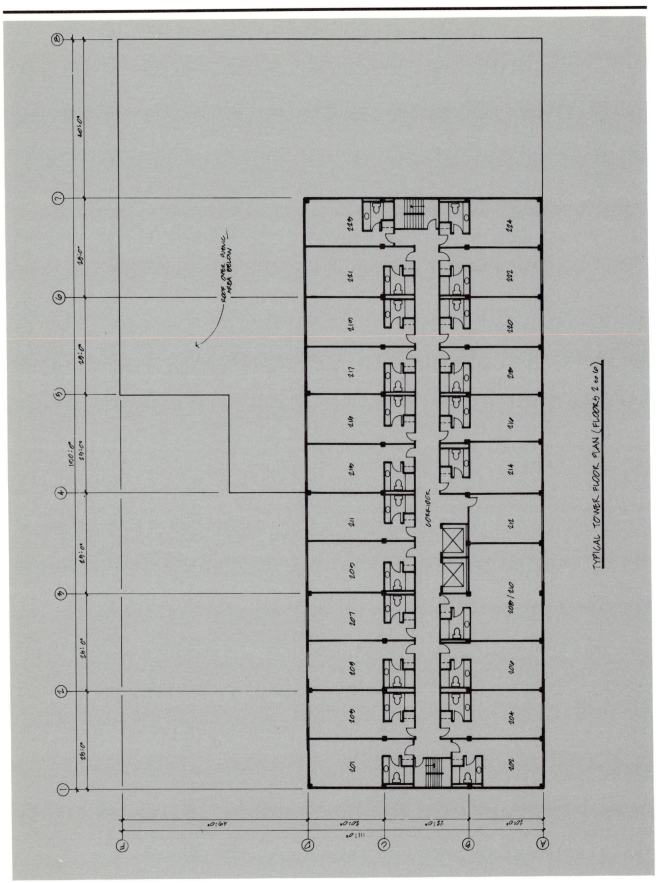

Figure 2.7

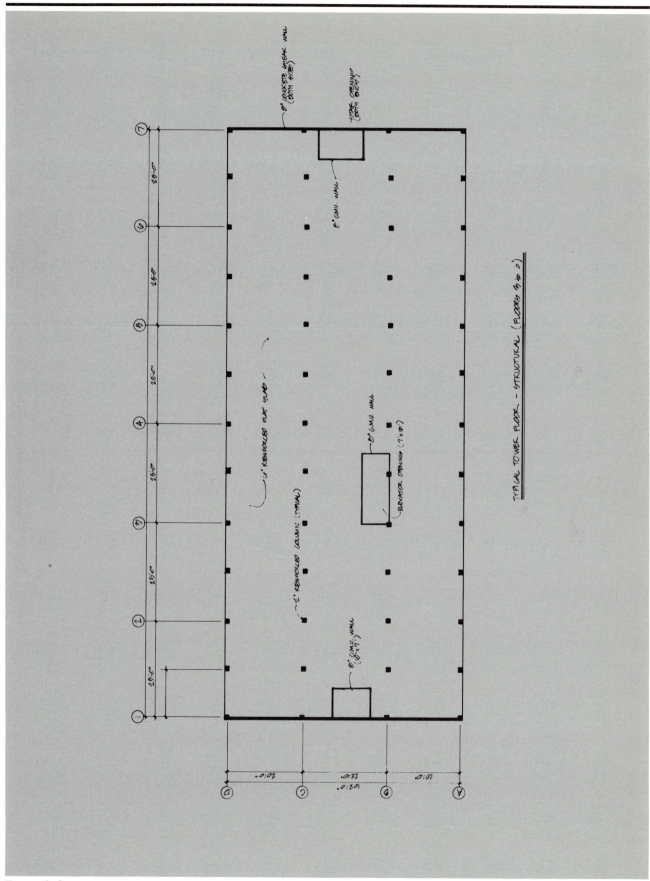

Figure 2.8

25

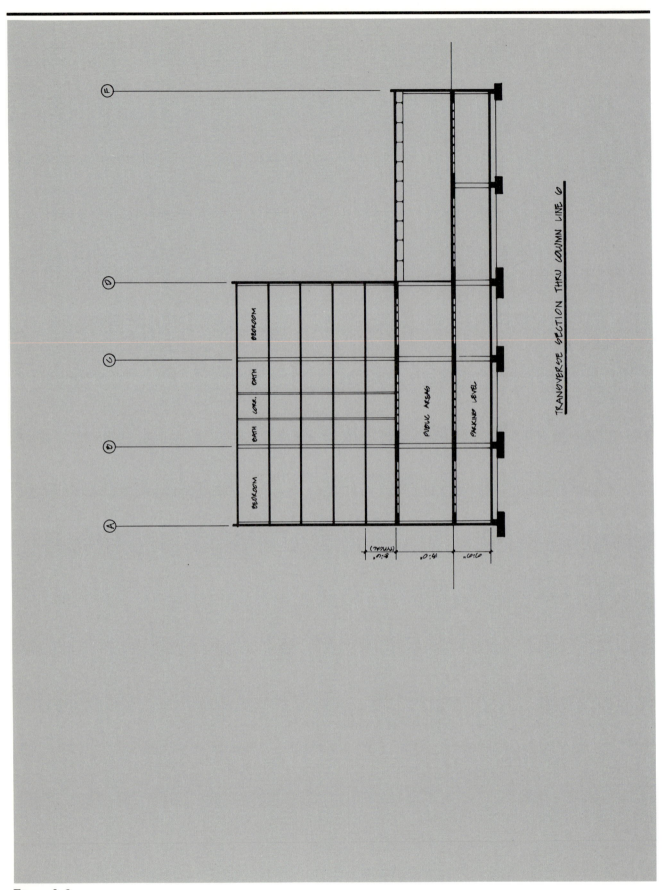

Figure 2.9

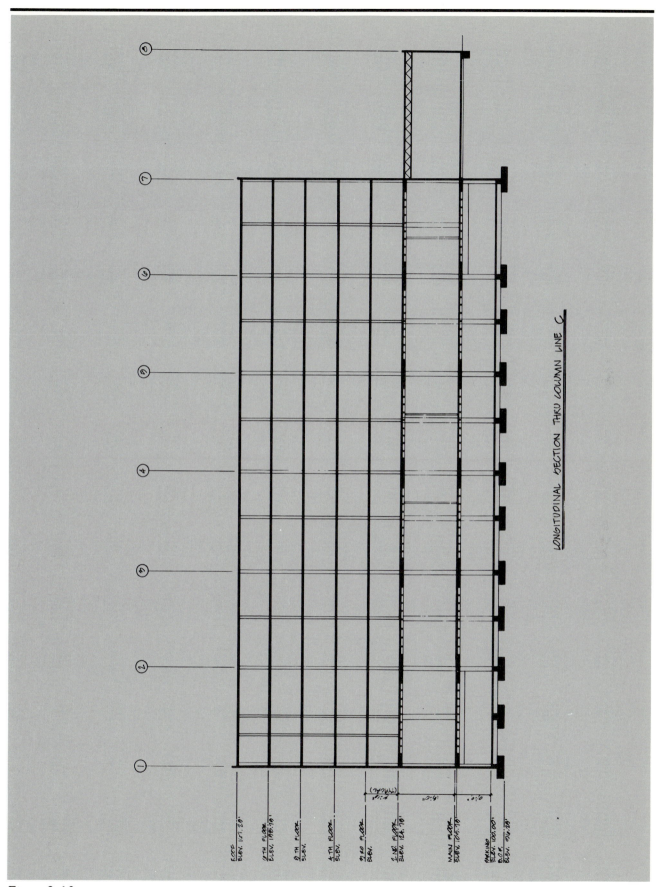

Figure 2.10

3

Pre-Construction
Planning

3

Pre-Construction Planning

After the contract has been awarded and the notice to proceed has been issued, the project manager must develop a project management plan. The best place to start this process is with a phase most accurately described as *pre-construction planning,* or *setting up for project control.* During this phase, the project manager establishes the basis for good job planning, and for efficient and effective monitoring as the job proceeds. Effective project control is an orderly process, which depends on accurate, reliable information distributed to project personnel. Setting the system up properly ensures effective performance from the beginning.

The project manager sets the tone for the job during this pre-planning phase. The quality of the initial set-up of a construction job has a profound effect on the entire project. If project personnel perceive a professional, well-thought-out approach, the job is more likely to proceed in the same manner. It is the project manager who is responsible for the job as a whole, and the entire project team looks to him for direction.

One of the major tasks of the project manager is coordinating all of the parties in the management process, and this, too, begins in the early planning stage. All should be encouraged to participate and share information about the project and its processes. Participants should also be made aware of their responsibilities — both during the set-up part of the process, and for regular procedures once the job is underway. Communications with project personnel involve more than simply recording in a manual the various requirements. It must instead be an active process in which the responsibility is assigned and acknowledged, and the tasks monitored to see that they are accomplished. There is no room for assumptions.

The project manager takes on the leadership role at this early stage, and maintains it throughout the job. He must communicate the requirements, and once the work is underway, perform follow-up procedures to ensure completion.

These early activities can be viewed as a twofold process, and will be treated as such in this chapter. First is the task of information gathering. We will cover possible and desirable sources of data, how the information may be best obtained, and who should undertake this task. The second part of the process is management — including establishing procedures. We will focus primarily on the ongoing decision-making, and on the record keeping needed to support project control. While other aspects of construction management, such as safety and personnel administration, are also very important, the emphasis of the discussion in this chapter is in keeping with the scope of this book, i.e., cost and schedule control.

The recommended procedures and checklist included here are not comprehensive, as it is not possible to cover every construction situation in one book. It is recommended that the individual project manager use these guidelines as a starting point, then expand and modify these procedures based on experience and the direction of his career. For example, the recommended file classifications could be used on a first job. During the course of the job, these categories might be expanded and culled appropriately, and the revised version used as the basis for starting the next, and presumably larger, job. Within a few years, a comprehensive file list will likely have been developed. The general guidelines presented here cover most project management situations.

Identifying Key Personnel

The first task that the project manager must undertake is to mobilize the company's resources. Part of this task is identifying key personnel. Depending on the size of the job, these individuals could range from a superintendent to a staff consisting of project manager, office engineer, scheduler, cost engineer, and secretary. Home office personnel who will not be assigned permanently to the job, but who will be responsible for specific tasks at particular stages of the project may also be part of the project team.

Providing Data to the Project Team

The next step is to note all sources of information for use by the project team. In these early stages, there are few sources of information that are available to everyone. Typically, all of the data is derived from the contract documents and estimate, plus the knowledge accumulated by the estimators during the bidding or procurement process. An excellent starting point is making sure that project personnel have access to the estimate data and contract documents. Ideally, copies of both would be issued to key team members. If this is too costly, copies of the documents should be provided in an easily accessible location and team members should be assigned the responsibility of familiarizing themselves with the contents of these documents.

Contract Document Review

A review of the contract documents should provide each project team member with an overall understanding of the job. The project team has a further opportunity to question any unclear aspects during a later meeting with the estimators. It is essential that the members of the project team have a clear understanding of the project. This is accomplished by simply spending as much time as necessary reviewing the plans, specifications, and estimate. During this document review, project team members should be looking for the following:

- **The size and scope of the project, including physical features of the structure.** The point here is that no one would start building from a set of foundation plans without knowing what was to be erected above the foundation.
- **Required construction procedures.** Contract documents do not typically describe specific construction procedures; each team member should visualize the process and then agree on the actual plan in the team meeting.

- **Special conditions and features.** This element should also be considered by individual team members and agreed upon at the project team meeting.
- **Physical limits of the site.** This category includes special construction problems that might be caused by the character of the site, by virtue of remoteness or surrounding structures.
- **Unusual situations.** The reviewer should look over the contract documents, taking note of those items which may cause unusual problems.

In addition to reviewing the plans and specifications, the team members should also review the general and special conditions. Information in this section should include:

- **Contract time allowed.** This figure sets a limit on the duration of the initial schedule. Any intermediate milestone dates that must be met, or portions of the project that must be delivered to the owner prior to the project's completion should be noted, as they clearly affect the project plan.
- **Owner's schedule, payments, and other administrative requirements.** Some owners, particularly public owners such as the U.S. Army Corps of Engineers, have specific requirements that must be met during the building process. These stipulations may have little to do with the actual construction. For example, pay invoices may have to be based on a critical path schedule rather than a traditional schedule of values. If there are specific requirements for the submission of schedules, it may be that only certain software packages will provide this capability. If the company does not already own a suitable system, it might first have to be procured. Setting up a new system creates purchasing and training requirements which must be considered.
- **Inspection and notification requirements.** Some contracts require a minimum notice for all inspections. Failing to take note of this requirement can result in otherwise avoidable delays during which work stands idle while waiting for the inspector to appear.
- **General administrative requirements.** Many owners require contractors to attend to administrative matters relating to public laws, financial institutions, etc. The contractor must set up appropriate procedures for those needs on the job site.
- **Claims requirements.** In today's legal climate, many owners are establishing time limits for submitting claims for extra compensation. The team should pay particular attention to the notice requirements. Certainly no contractor wants to be in the position of losing an otherwise valid claim because of having failed to meet these requirements.

Estimate Review/ Estimator Meeting

After reviewing the contract documents, the team members should review the estimate thoroughly. The project manager should then schedule and conduct a meeting with the estimators, allowing the team an opportunity to ask questions. This meeting should be one of the earliest, and certainly before any meetings with "outsiders," such as subcontractors or owners.

In the early stages of a project, right after the bid is awarded, the estimators and their documents are probably the best and most complete source of information in the company about a given project. It is not enough, however, to simply read the estimate and look at the numbers. The information in an estimate is not arranged to help the actual management of a project. It is arranged to help make the bid process accurate and efficient.

As such, it may need some interpretation by the originating estimators to make it useful to project management. A meeting between the estimators and the project team insures that the assumptions built into the estimate are reflected in the project plans.

In reviewing the estimate and questioning the estimators, the project team should look for the following.

Special Conditions and Features   In addition to the special conditions and features picked up by the project management personnel, the estimators may be aware of other situations which are not readily apparent in the estimate documentation. It should be the estimator's responsibility to pass this information on to the field personnel, but it cannot be assumed that this will be the case. The estimator may well be deeply involved in preparing another bid, so members of the field team must ask specific questions.

Assumptions and Limitations   Even in straightforward projects without special features, the estimator will have made assumptions about how the project is to be built. Typically, these assumptions are made in the interest of getting the bid put together on time. While such assumptions are usually valid, it should be noted that the project team is not obliged to assume that they are completely infallible. In fact, the management team should be encouraged to find a better solution to problems of productions. Any improvement, after the bid is won, results in pure profit, not money left on the table. At a very minimum, questions should be asked about the following areas:

- **Weather.** Weather can be thought of in terms of two different factors: location and time of year. Location is clearly not alterable, but the time of year may be flexible. For example, while the use of overtime is usually not a good idea, there may be cases where the shrewd application of overtime to structural work in good weather may permit indoor work to take place in winter weather in an enclosed building. A "normal" schedule, on the other hand, would simply mean waiting for the spring before resuming structural work.
- **Crew composition.** The management team clearly needs to know what crew make-ups were assumed in planning both the unit costs and the scheduled time in the estimate. Initial crew assumptions serve as a starting point in the scheduling process, but should not be taken as absolutely reliable without checking. It is possible that the estimator has allowed too few crews, or crews which are too small.
- **Equipment.** It is also necessary to determine what equipment has been assumed, particularly since equipment has such an impact on other parts of the job. Estimates often contain equipment assumptions based on a particular piece of gear being on the job for a period of time, rather than tying the equipment to the cost of specific tasks. This approach is especially common with large pieces of equipment such as tower cranes. Problems can arise if the equipment is either too small to handle all that is expected of it, or if the time allowed on the site is not adequate to do all the jobs for which it is intended. Equipment choices also have an impact on crew assumptions in that the time allowed for a piece of equipment on site may be predicated on a crew moving at a given rate; if the crew is too small to move at that rate, then clearly, the size of the crew or the time frame for the equipment must be altered. The estimator may also have made certain assumptions about renting, leasing, or buying. Such decisions may be governed by company policy (i.e., if the company owns

it, we use it). Again, the project manager may not be obliged to use the estimator's plan, but it does serve as a starting point for the actual decisions as to equipment use.

Note: the project manager should be aware of any deals made by the estimator with the subcontractor regarding the use of general contractor-owned equipment. This issue can be very complex, particularly on high-rise jobs where only a limited amount of equipment can fit onto the site at any one time.

- **Location.** As previously mentioned, the location of a site can have a great influence on construction cost and schedule. The two types of sites that tend to cause the greatest problems are those that are either isolated or tightly restricted. Both tend to increase production costs and schedule time, by virtue of the difficulty and expense associated with getting material and/or labor into the site.
- **Logistics and procurement.** While site location is an important factor in logistics, it is not the only one. In general, any situation that requires materials or equipment that are subject to manufacturing and delivery problems requires special attention. The estimator should be questioned on this point since, hopefully during the process of compiling the bids, he will have identified the items to be installed which will take longer than normal to be delivered. Once these items have been identified by the project team, they can be properly factored into the schedule. Tracking these items during the job is also extremely important. Promised delivery dates are seldom moved up; they are much more likely to be put off, and during construction, adjustments will have to be made.
- **Subcontractors.** Of all the items that must be discussed with the estimator, subcontract arrangements are among the most important. The scope of work of each subcontractor is never clear immediately after a bid opening. The time pressures of bid day are simply too great to permit precise definition of scope at that time. Also, the documentation of the subcontractor bidding, including all the call-ins, bid alterations, and price cuts, are invariably confusing and require explanation by the estimator. In particular, the project manager will be responsible for executing the buy-out with the subcontractors as the job proceeds. He or she must know the conditions under which bids were tendered by the subs. The project manager must also know the identity of each subcontractor, the scope of work for which the subcontractor bid, and the price quoted. The project manager should also be made aware of any special contractual exclusions made by the subcontractors, and any long-lead-time items within the subcontractor's scope of work.
- **Quantity Takeoffs.** Finally, the project team should inquire as to the nature of the material takeoffs done by the estimators, and should understand the format thoroughly. This part of the estimate serves as a handy source of information for the field production planning and scheduling which will occur as the job progresses. These original quantity surveys will probably have to be reworked in part, but they provide a good starting point.

Initial Project Team Meeting

After reviewing the contract documents and the estimate, and after meeting with the estimators, the project manager should meet with the project team. The purpose of this meeting is to set goals for accomplishing the remainder of the pre-job planning tasks, and to assign responsibility for getting those tasks done. Such a meeting also provides an opportunity to solicit ideas and comments from the project team members, both for building the job and for starting the ongoing project control procedures. Ideally, there should be a corporate operations manual in existence to guide new personnel. If there is no manual, one should be created and should include the above-mentioned planning procedures.

During the initial meeting, the team should consider the following:

- **Overall job approach.** It is very rare that a company gets a job of a type that no one in the company has seen before. In fact, it makes sense to assign most of the personnel to a job on the basis of their having done that kind of work before. The team should therefore be able to decide fairly quickly on the primary methods of construction. In the event several possibilities exist, an appropriately knowledgeable team member should be assigned the task of investigating the alternatives and reporting the results to the project manager.
- **Most critical parts of the job.** In most types construction, the most difficult or likely trouble areas can be identified by the team members based on past experience. All members of the team should be made aware of these areas. By getting this information out onto the table, decisions can be made as to the best course of action, which will result in the most effective construction sequence.
- **Project control needs.** Since the intent of the meetings is to set up the most effective project management process possible, the team will also be considering the control system to be implemented. While many of the systems, such as cost, may be dictated by company-wide requirements, some, such as scheduling, are up to the team. The size and character of scheduling systems to fit specific jobs are discussed later in this book, but it is worth noting here that during these initial meetings, the team and the project manager must start to consider how scheduling is to be implemented. Scheduling choices are based, primarily, on the degree of detail desired, the way in which the information is to be displayed, and how often that information is needed. Also, the responsibility must be assigned for maintaining the scheduling system.
- **Project administration.** Complete coverage of this subject is beyond the scope of this book. However, project administration is an important topic that should be discussed in these early meetings when the project team sets up procedures. These procedures will be determined by such diverse needs as corporate policy, the owner's requirements, and local ordinances.

  It is important at this point to assign the responsibility for creating rough drafts of major subcontracts. This is done in anticipation of the buy-out meeting which will fix the scope of work and the price, drafting the major material purchase orders for mailing to suppliers, and last but not least, identification of submittal data requirements by job items.
- **Remaining pre-job planning.** Finally, the team should plan the actions and assign the responsibilities for dealing with the pre-job planning which must occur with agencies and parties outside the company. These parties and items will be discussed in the next section.

Pre-Planning with Other Parties

As is the case with employees, subcontractors and other "outsiders" respond well to professional management and respond poorly to sloppy procedures. The project team should use the same professional approach in dealing with these other parties to the contract. While there are many similarities, each outside party has its own peculiar set of requirements for the most effective construction process.

Subcontractors

According to some experts, the most challenging aspect of construction today is the effort to achieve effective, on-time use of subcontractors on job sites. While the subcontracting process does relieve the general or prime contractor of much of the risk stemming from labor cost overruns, the problems associated with scheduling and actually getting the work accomplished are, in some ways, worse. The reasons for this situation are many. One problem is that each subcontractor is a separate corporate entity with obligations beyond the job site; few are on the job site full-time. Early, effective pre-planning can do much to prevent the problems associated with extensive subcontracting.

The general process for dealing with subcontractors can be described in the following three steps:

1. Setting up the contract, in a process which is known as "buy-out."
2. Involving the subcontractor extensively in the initial planning and scheduling process.
3. Involving the subcontractor in the monitoring and any subsequent re-planning processes that may occur.

The first of these phases is discussed in the following paragraphs; the last two are covered in detail in subsequent chapters.

The buy-out is the process in which the general contractor makes the final subcontractor selections, negotiates scope of work and price, settles contractual terms, and signs the subcontract. Also considered at this time are initial scheduling concerns, schedule meeting and notice requirements, subcontractor representation in dealings with the owner, and the submittal data and equipment delivery issues associated with the subcontractor's work.

The first of these concerns is the definition of scope of work and price. As previously noted, the bidding process leaves no time for adequate definition of scope of work. This task is best accomplished later in meetings between general and subcontractor, at which all of the items of work are very carefully defined. Defining scope of work and price is a task involving great amounts of detail; it is a job that is best done one piece at a time. A set of drawings and specifications is used, on which each item of work for which the subcontractor is responsible is marked in color. From the drawings and specifications, a list is developed which both parties can keep as a part of the contract. This list should cover both included and excluded items, and should be as specific as possible. While this method of marking up drawings and specifications is not the only way to conduct a buy-out meeting, the purpose of the process is to remove all possible ambiguity from the subcontract. One of the benefits of creating a well defined set of contractual obligations for the subcontractor (which will in itself prevent future claims), is the fact that it is probably not possible to talk about *what* will be done without talking about *how* it will be done.

At this time, the project manager and subcontractor should also address scheduling issues. As the activities are defined, a commitment must be obtained from each sub regarding the duration of each activity. The submittal data is required before an activity can begin, and it must be known what material must be ordered and delivered prior to the start. These items should be included in the list of contractual obligations for the subcontractor. The information will be used in the development of the critical path scheduling sequence and times.

The subcontractor must also be made aware of the ongoing scheduling requirements for all parties on the job. The basic idea is that the project manager must conduct regularly scheduled meetings during the course of the work; attendance and participation will be required of all parties. Despite the fact that many contractors and subcontractors consider meetings a waste of time, experience in the industry has shown that well-conducted, regularly scheduled meetings, in which subcontractor ideas are encouraged and used, and subcontractor commitment is required, are the most effective means of achieving the real coordination which is so vital to successful job progress.

Suppliers

In addition to the subcontractors, the major material suppliers must also be contacted and the information they contribute worked into the planning process. During the bidding, suppliers, like subcontractors, are working under the pressures of time, and while their quotes typically contain terms and dates of delivery, it is always best to confirm that data. The project manager cannot assume that the estimate contains information on all items that have the potential to delay the schedule. For example, because of time pressures, the estimator may commonly use average, current prices for a bid submission, rather than specific quotes. There will also be no information gathered about delivery times at this stage.

The specifications and plans must be thoroughly reviewed for the following purposes: to identify all articles of material which must be ordered and delivered, to identify and contact suppliers, to confirm prices, and to obtain delivery commitments. This information must then be incorporated into the budget and schedule.

Even materials that are not difficult to obtain must be reviewed in terms of their effect on the scheduling process. Many owners, especially public agencies, require submittal data on even the most common of materials, such as lumber and concrete. These submittal data requirements can affect the project schedule if the submissions and subsequent approvals are not carried out on time.

The purpose of the pre-planning process is to identify and assign responsibility for the activities of the job. Effective pre-planning prevents the problems associated with making purchases just as installation is due to begin. The administrative task of setting up logs and continually tracking the status of these items must also begin at this point.

Owners and Their Representatives

One of the universal characteristics of construction owners is their desire to monitor how their project is being built and how their money is being spent. Most of the owner's attention is, understandably, directed toward ensuring good quality in the finished product. Owner involvement frequently creates administrative work for the contractor in the form of progress reports and additional record keeping requirements at the site, such as insurance confirmations and lien releases. The vast majority of these requirements are clearly designed to protect the owner's interests.

Many owner requirements are covered in the contract documents and will therefore have been discovered during the document review. However, a pre-job meeting between owner and contractor will almost always be required. It is at this meeting that the contractor can discover the "unwritten rules" and clarify issues which may not be totally clear in the contract documents. This meeting will also be the start of personal working relationships between the owner's and contractor's representatives. These contacts can sometimes degenerate in an atmosphere of acrimony and distrust. It is therefore important to start off and remain on as positive and professional a footing as possible. A contractor must recognize the fact that the owner's concerns and fears are legitimate, since a large amount of money is typically at stake. The contractor should be properly organized for the initial meeting with the project team, and display a cooperative attitude toward the information needs of the owner's representatives.

During the initial meeting, the following topics are of primary concern to the project manager. First, make sure that any items not clearly expressed in the contract documents now become known and understood. Very often a few words in a contract document will not even begin to describe a potentially complex situation. For example, it is common for a contractor to perform work while an owner continues to operate a business in the same structure. Under these circumstances, the contractor must know how to avoid damage to existing equipment and work, and the procedure for coordinating the various activities of construction with the closing or opening of various parts of the building. There may even be other contractors who have been hired by the owner and who are working at the facility at the same time. This situation usually requires frequent and effective communication among all concerned parties for the life of the project. First meetings with the owner set the tone for this aspect of the project as well.

There are many other issues that must be coordinated with the owner in the initial stages. For example, most construction projects require temporary utilities. These are often coordinated with the utility companies through the owner. Frequently, there is also the matter of owner-furnished materials or equipment which must be identified, delivery dates verified, and installation details confirmed. It should also be noted that the contractor must inform the owner of the specific, potential impacts on the schedule if delivery of owner-furnished materials or equipment are not made on time.

Public and Government Agencies

Almost all construction projects must meet the requirements of a regulatory government body. It has traditionally been the contractor's responsibility to acquire permits for the construction process, but that requirement has become much more complex in recent years. For example, many permits, such as environmental approvals, must be obtained by the owner prior to the start of design, with an additional permit for actual construction obtained by the contractor at the start of work. Some permits are predicated on the issuance of other permits; this whole issue has the potential to become a complex and difficult business.

There are several actions that the project manager can take in order to cope effectively with permits. First, a comprehensive list of required permits and the parties involved should be made. This list must include: all the parties involved with individual permits, identification of those responsible for obtaining and paying for the permit, and the requirements for obtaining the permit. This list should be carefully cross-checked and coordinated with the owner and subcontractors.

Most construction contracts assign the contractor the responsibility of complying with local laws without saying what those laws are. The project team must therefore do some careful research in order to determine exactly which local laws and ordinances require the contractor's compliance. The project team can get this information by going to the local building department and environmental agencies and obtaining literature on the local requirements. It may be helpful to establish personal contact with the officials involved after reading the department's literature. This approach makes it easier to find out about any hidden rules and agendas.

Unions and Labor Suppliers

Finally, the contractor and project manager must be aware of the procedures, problems, and pitfalls of obtaining labor in the location of the job. Basic information about labor availability and rates should have been known at the time of the bid. The project manager will, however, have to follow up in order to find out what the actual procedures are for hiring in a designated area. Aside from the more obvious source — local business agents and other union officials — it is a good idea to contact local contractor associations and government employment development agencies.

Setting Up Management Procedures

Concurrent with the information gathering process, the project manager must also begin setting up management procedures for the project. Clearly, some of these procedures will be influenced by the specific requirements of a particular project. Most, however, will be similar in nature from project to project, especially for a series of projects performed for the same owner, or in the same local area. Some basic guidelines for these procedures follow.

Assigning Responsibility

The assignment of responsibility is a task that continues throughout the project. It is, however, often neglected in the early stages. As a result, many important tasks do not start off as regular, monitored events, and are consequently discovered not done at a critical time.

The basic concerns in assigning responsibility are as follows: First, make sure that every ongoing task is identified — especially those related to cost and schedule. Second, a competent individual must be assigned to monitor each task on a regular basis. For example, in scheduling, who is to track the progress of each activity on a periodic basis, and how often is the tracking to be done? Or, in the case of daily labor time cards used to record the expenditure of labor man-hours, who is to fill the card out? Who checks it, and who transmits it to the home office for processing? The project manager must follow up to see that the responsibility assigned is in fact assumed by the assigned individual.

Coordination on the Job Site

Construction projects are complex endeavors, and procedures must be established to ensure effective communication among all parties involved in the project. Communication requires both informing and listening to all participants. For example, schedule obligations should be clearly explained to subcontractors. In fairness and in the interest of maintaining productive harmony on the job, the subcontractors' concerns must also be heard prior to decisions being made about their assignments. The means of communication can take many forms, but consist primarily of: 1) regular meetings of the concerned participants, and 2) recording and distributing the results of the meetings in written form.

The first element, regular meetings, must be made productive by the project manager. Otherwise, the other parties will simply come to regard the meetings as a farce, with no attention being paid to the results. To use scheduling as an example, a regular meeting should be held every week. In this meeting, the schedule for the previous week and upcoming two weeks would be discussed. The project manager must ensure that an up-to-date schedule and agenda are provided to all parties required to attend, and that the discussion is productive and to the point.

Following the discussion, the project manager should announce his decisions, basing the schedule he proposes on the commitments given publicly by each subcontractor. The new schedule should then be distributed in written form within one day.

In holding schedule meetings on this basis, the project manager invites participation by the subs, which helps to establish an overall team commitment. He also obtains "public" commitment from the subs, which is more likely to be met than a privately made promise. By distributing the results promptly, the project manager provides written documentation for the job, and re-confirms in everyone's mind the decisions made. Effective use of the project meeting is one of the project manager's most powerful tools.

Coordination with Other Parts of the Company

In addition to coordination of the subs and other on-site parties, most jobs also require some degree of coordination with the home office. These procedures are typically less demanding than those directly related to the construction work, but are no less important. For example, all companies have a set of procedures for accounting and payroll. These procedures are company-wide and must be dealt with at each job site. Ideally, they are defined in a company operations manual or guide, and need not be re-invented for each new project. If such a manual does not exist, then company efficiency would be well served by creating one.

Setting Up a Good Record Keeping System

Finally, in order to control the many documents generated by a typical job, the project manager must see to it that a consistent and workable record keeping system is set up and maintained on the job site. A job site filing system must be set up, and responsibility must be assigned for keeping each file. The job secretary may not be the most logical person to keep all of the files. For instance, working files, such as those for scheduling, might logically be the responsibility of the scheduler or other individual who is responsible for maintaining the schedule. Again, record keeping should be consistent throughout the company, but in the event it is not, the following filing breakdown is recommended as a starting point.

A. **Correspondence**
   One file for incoming, one for outgoing, arranged chronologically.
   1. Owner/Client
   2. Architects/Engineers
   3. Subcontractors (by specification division)
   4. Regulatory & Inspection Agencies
   5. Suppliers (by specification division)
   6. Miscellaneous

B. **Transmittal Letters**
   One file for incoming, one for outgoing, arranged chronologically.
   1. Owner/Client
   2. Architects/Engineers
   3. Subcontractors (by specification division)
   4. Government & Inspection Agencies
   5. Suppliers (by specification division)
   6. Miscellaneous

C. **Meeting Notes and Minutes**
   Arranged chronologically.
   1. Regularly Scheduled:
      a. Weekly Job Conference
      b. Weekly Subcontractor Scheduling Meeting
      c. Other

  2. Non-Regularly Scheduled:
   a. Owner/Client
   b. Designers
D. Contracts
  1. Owner/Prime
   a. Main Contract
   b. Addenda (by number)
  2. Prime/Subcontractors
   (by specification subdivision)
  3. Outside Consultants
   a. Surveyor
   b. Form Design Consultant
   c. Other
E. Changes and Scope of Work
  1. Specifications
  2. Requests for Information
   (consecutively number all R.F.I.'s)
  3. Responses to R.F.I.'s (by number)
  4. Bulletins/Design Changes
  5. No Cost Change Orders/Field Directives
  6. Change Orders Pending
  7. Approved Changes
  8. Unapproved Changes
  9. Claims in Process
  10. Delays/Time Extensions
F. Project Control
  1. Cost/Budget
   a. Original Estimate
   b. Bid Analysis
   c. Budget
   d. Weekly Labor Cost Summaries
   e. Weekly Equipment Cost Summaries
   f. Monthly Material Cost Summaries
   g. Monthly Job Cost Summaries
   h. Other Regular Cost Reports
  2. Schedule
   a. Activity Work Sheets
   b. Networks
   c. Periodic Schedule Reports & Runs
    (arranged chronologically)
    1. Planning Schedules
    2. Bar Charts
    3. Other Runs
G. Invoices and Payments
  1. Schedule of Values
  2. Invoices to Owner
  3. Payments Received
  4. Invoices from Subs (by specification section)
  5. Payments Made to Subs (by specification section)
  6. Invoices from Suppliers (by vendor)
  7. Payments Made to Suppliers (by vendor)

H. Reports
Arrange chronologically.
1. Daily Job Logs
    a. Project Manager
    b. Superintendent
    c. Project Engineer
    d. Other
2. Telephone Logs
3. Progress Photos
4. Weekly/Monthly Job Progress Reports
5. Change Order Status Reports (by change order number)
6. Submittal Data Status Reports
    (by specification section and by subcontractor and supplier)
7. Material Delivery Status Reports
    (by specification section)
I. Tests & Inspections Reports
1. Inspection Logs
    (by specification section)
2. Inspection Reports
    (by inspecting agencies and companies and by specification section)
3. Test Reports
    (by inspecting agencies or company and by specification section)
J. Material Procurement and Deliveries
1. Master Material Order List
    (by specification section)
2. Material Order Status List
    (by specification section)
3. Purchase Orders
    (by specification section)
K. Submittal Data/Shop Drawing Logs
    (by specification section and by subcontractors and suppliers)

4

Introduction to Scheduling

Introduction to Scheduling

As noted in Chapter 1, it is normal practice in the construction industry to maintain separate systems for time and cost control. The next several chapters of this text specifically address the subject of time control, including the step-by-step process of setting up scheduling procedures. Before proceeding with the specifics, however, it is useful to look at some basic facts and ideas about the process of scheduling and how it works.

The Critical Path Method

The emphasis of this book is on the use of critical path method (CPM) techniques of construction scheduling. Of all the techniques available to a project manager, CPM has proven to be the most useful and effective means of developing and displaying the information needed to control the time variables on today's job sites. The basic CPM technique was developed in the late 1950's, primarily for the purpose of controlling large manufacturing and construction projects. It has been further developed and refined since, and has evolved into a tool that is well suited to the construction process.

Most people familiar with the construction process recognize the fact that a project is composed of tasks which are separate, yet interdependent. For example, in building a house, both foundation and stud wall are essential elements. The crew that forms and places the foundation is very different from the one that erects the stud walls. Nevertheless, these tasks are interdependent in that the walls cannot begin until the foundation is complete. The most difficult task in construction is keeping track, and deciding the correct order and timing, of a large number of these individual, yet interrelated tasks; the critical path method of scheduling addresses this issue.

The CPM technique is simpler and more flexible than it might first appear. It takes the building process one step at a time and separates the project into workable sub-parts or activities. A plan is made for each activity to be performed in the correct sequence. The task of scheduling becomes a systematic, one-piece-at-a-time endeavor.

The basic steps, or phases, of scheduling are as follows.
A. Planning
B. Scheduling
C. Monitoring and Controlling

Each of these phases has sub-steps. The first phase, planning, involves:

1. Breaking the project down into workable sub-tasks, commonly called *activities.*
2. Deciding the order in which these activities are to be performed.

The result of the planning phase is a *logic diagram,* or *network,* which is an initial graphic representation of a plan of what to do and the order in which to do it. This phase of the process is illustrated in Chapter 5.

The second phase, *scheduling,* adds a time element to the planning phase; the sub-steps for this phase are:

3. Determining a reasonable duration for each individual activity.
4. Calculating the duration of the project as a whole.

The product of this second phase is a series of time plans, typically presented as *planning schedules,* or *bar charts.* This type of display is shown in Chapter 6 and in Appendix A.

The last phase, *monitoring and controlling,* consists of:

5. Measuring the progress of the project.
6. Comparing the actual progress against the schedule developed during the scheduling phase.
7. Taking corrective action if the actual progress deviates significantly from the schedule.

The monitoring and controlling phase is covered in detail in Chapter 9.

Learning CPM Techniques

Many people who have tried to implement the critical path method have found it a difficult task. Most of this difficulty stems from not recognizing the basic simplicity of the process, and from being overwhelmed by the "gurus" of scheduling who have made the process seem far more complex than it really is. Much of this book is devoted to straightforward, workable techniques for CPM scheduling. As these techniques are presented, it is helpful to keep in mind the following general guidelines.

First, take the process one step at a time. The chapters of this book are presented in the same, step-by-step manner, outlining each of the major tasks that must be performed in order to achieve effective construction project management. These tasks and methods are discussed in the actual order in which they would occur on a project.

Second, recognize that the CPM technique is a way of representing what a manager intends to do; it does not require that the manager build in an unfamiliar way. Modern CPM techniques and software systems have more than enough flexibility to represent virtually any possible plan of action desired, so there is no reason to have the "tail wag the dog."

Third, the project manager and other project personnel should recognize that using CPM effectively will require an investment of time and energy on their part. Using CPM requires skill, and no skill can be developed without some effort in learning it. To use an analogy, no one in construction would expect to use a new laser surveying system without some training and practice. CPM is like the new laser surveying system in that it is a better productivity tool, but requires an initial investment of time as well as money. CPM is also like the new surveying system in that no contractor is likely to continue with an old tool when his competitors are using a new and more productive one.

Fourth, developing schedules is a creative process. It is analogous to an architect developing a set of plans for a building. No one ever created a set of working drawings without first going through a lot of sketches on tracing paper, then schematic and design development drawings, and finally, working drawings. Those developing a CPM schedule should be prepared to do a lot of erasing and rewriting as decisions are made and altered, and the plan of action develops.

Potential Pitfalls of Using CPM

First and foremost among CPM's potential pitfalls are those related to the human element. It is very common, for example, for a contractor to require the use of CPM for scheduling projects, and to use schedules that are developed and presented by professional schedulers. These professional schedulers have no stake in the outcome of the project, and may fail to consult the project manager and other project personnel. The result is a schedule which frequently does not reflect reality, and certainly not the project manager's reality. In this situation, those who are responsible for the performance of the project are, in effect, being told how to run the job, and may, quite understandably, be resentful. As a result, the schedule may be largely ignored by field personnel and become useless as a monitoring tool. In this case, the job as a whole suffers and the management of the company is unable to monitor progress until it is too late to correct the overruns.

The best way to avoid an unrealistic schedule is to provide project management with the tools and training to develop and use CPM effectively, and for top management to require its use in tracking and reporting progress. This is the ideal, in which every field manager regards good scheduling as an essential part of his or her job, and has the skills to use it properly. In the real world, the company's management must be sure that any schedules developed by others reflect the thinking of the field personnel, and that field personnel see the schedulers and the schedule as serving them and not as an imposed duty. If the schedule does not reflect the thinking of the people who have to live with it, it will not have their commitment, and cannot be effective.

Second among potential pitfalls is over-complexity. This problem can usually be identified when the schedule reports tend to gather dust rather than fingerprints. The schedule may be particularly susceptible to this problem when a "sophisticated" computer system is used. The problem brought on by complexity is that the schedule does not serve the project managers, but rather becomes an end in itself, and is ultimately ignored. The computer's capacity for generating large amounts of paper seems to be very difficult to resist and quite often one or more of the following occurs.

- Huge volumes of reports are generated which are so complex and bulky that it is difficult to read them all, much less pick out the important ones;
- The reports are confusing or are in an inappropriate format, and do not concentrate on the problems at hand, i.e., they do not promote management by exception;
- The project manager is flooded with detailed reports when, in fact, summary reports are needed; or vice versa;
- There is a severe lack of flexibility in the reports or in the schedule itself.

5

Planning the
Project

Planning the Project

As noted in Chapter 4, the first major step in the scheduling process is *planning,* which consists of:

1. breaking the job down into "sub-tasks," and
2. establishing the sequence of work.

This chapter describes each of these parts of the process in detail, and shows how the two are linked.

Breaking the Job Down into Activities

The first part of the scheduling process, breaking the job down into sub-parts, is probably the easiest. In fact, anyone who has worked in construction is probably aware that a foreman, superintendent, or project manager performs this task, either intuitively or deliberately, as a normal part of managing. Construction managers typically think of a job as a series of distinct steps, separate from one another. This separation process begins in the estimating stage, and continues through the hiring and scheduling of subcontractors, and in the setting up and directing of construction crews.

In using CPM techniques, the separation of sub-parts is more formal and more complete. There is no set way to go about this subdividing process, but there are some guidelines which can help to ensure that activities are not overlooked or ignored.

Listing the parts of the job can begin very early in the life of a project. For example, during the reviews of plans and specifications, a construction manager will probably already be thinking of the different tasks that have to be performed. The subdivision process is best begun using the basic construction documents, the plans and specifications, and then expanded based on additional information from other sources.

General Activity Types

The easiest and most general way to view the task breakdown is by recognizing all activities as falling into one or more of the following categories.

1. **Production Activities** — directly related to construction, involving crews, materials, and installing elements of the building. Examples are "erect steel studs," or "place foundation concrete."
2. **Procurement Activities** — required to order, purchase, and ensure delivery of the materials and equipment that are to be used in erecting the

project. Examples are "order condenser units," "fabricate storefront frames and glass," or "deliver hollow metal door frames."

3. **Administrative and Support Activities** — "secondary" to the construction process, but vitally important in the complex and legally-oriented environment that exists today. Examples are "submit water fountain shop drawings," or "approve steel shop drawings."

Every element installed in a project has an administrative process which must take place before actual construction can begin. To illustrate this idea, visualize the steps that must be performed in order to get structural steel for the sample hotel project in place. These steps might include:

- Prepare structural steel shop drawings
- Review structural steel shop drawings
- Order structural steel
- Fabricate structural steel
- Deliver structural steel
- Erect structural steel

The first two steps are administrative in nature, the next three are procurement, and the last is actual construction. The reader will note that of these six steps, only one actually involves crews and construction on the job site. Nevertheless, all are necessary for the installation of structural steel.

Specific Activity Types

After breaking the job down into general activities, a manager or scheduler working up a construction schedule further divides these categories into more specific items. This classification process might be based on the following features:

Physical Elements of the Project   This is the most basic of the groupings and the one that comes to mind most readily. In fact, anyone who works in construction tends naturally to view a project in these terms.

In looking at the sample hotel project, it would be logical to think in terms of "foundations — garage area," and "columns — garage area." Clearly, each of these two parts of the building must be scheduled with different formwork, and in all likelihood, different crews will perform the work. Also, it is clear that the columns rest on the foundation and must therefore be installed after the foundations are complete. All of these factors make it necessary to separate the two activities.

Trade, Skill, or Crew Involved   The various trades used to erect the parts of the building must be directed by different foremen, and are therefore required to be kept separate in any schedule. Again looking at the hotel project, typical activities might be, "form concrete retaining walls," "erect masonry walls," "erect structural steel," and "erect metal stud walls." Each of these activities involves a different trade or skill — carpenters, masons, and ironworkers, respectively. Each activity is managed and directed separately, and should therefore be kept separate in the schedule.

Contractual Divisions   In today's construction projects, increasing amounts of the total work are subcontracted. Each subcontractor is a separate corporate entity, with his own set of contractual obligations to the general contractor. There are several reasons why each subcontractor should be kept separate on the schedule.

First, each subcontractor is managed and scheduled for work individually. The electrical subcontractor will probably be scheduled to rough-in after the heating, ventilating, and air-conditioning subcontractor.

Examples of this type of subdivision on the sample hotel might be:

Rough-in electrical — main floor
Rough-in ductwork — main floor
Install storefront — gift shop

The second reason for keeping the subcontractors separate is that payment to each depends on progress. It is generally regarded as good practice to clearly separate one subcontractor's progress from that of another. In this way, disputes over the responsibility for lack of progress are prevented. In the event a dispute does occur, separation of the subcontractors helps the general contractor accurately assess who is at fault. The process of claims resolution is also greatly simplified if the schedule keeps all contractors separate — an important consideration in today's legal climate.

Organizational Responsibility   This idea of subdividing also applies to the organizational subdivisions set up by the general contractor. For example, if a project is large enough to require two field superintendents, the activities for which each is responsible should be kept separate. Or, if a different part of the company is involved (such as a separate formwork design department which was part of the home office), then the task of "design formwork — tower slabs" should be kept and tracked separately from the field work.

Physical or Geographic Area   Often, if a job is large enough, it might be organized along what could be called *geographic lines*. This practice is very common in such projects as petroleum process plants, which are spread out horizontally over large tracts of land. In these cases, it is important to be able to track progress across the site. The principle can be applied to a project as small as the sample hotel. For instance, the two roof areas might well be installed at separate times, leading to two separate activities, such as:

Install roofing — dining and kitchen area
Install roofing — tower

In the case of vertically-oriented buildings, it is important to be able to track progress *up* the building by area. A manager needs to see activities organized in this way, since it is very likely that similar crews could end up working at several levels at the same time, for example:

Install metal studs — main floor,
Install metal studs — 2nd floor, tower,
Install metal studs — 3rd floor, tower, etc.

Level of Detail

How much detail should be provided in developing the lists of activities? This concept is discussed at length in Chapter 8, in dealing with coding activities for retrieving information. The basic idea is that the more tightly one wishes to control a project, the more detail must be contained in the schedule. To illustrate this point, consider the following two lists of activities for forming a tower floor in the sample hotel. The first list contains a high level of detail.

Form columns, east end
Form columns, west end
Form shear walls, east end
Form shear walls, west end
Form elevator walls, center
Form floor slab, east end
Form floor slab, west end

The second list is a much lower level of detail, as follows:

Form columns
Form shear and elevator walls
Form floor slabs

The degree of detail required by the project manager depends on a number of factors, not all of which may be apparent at the early stages of schedule development. One basic consideration is the level of control desired. If tight scheduling of concrete crews and formwork is necessary, a higher level of scheduling detail may be needed. In this case, the manager may wish to bring in ductwork crews when only one-half of the floor is complete in order to get an early start on rough-ins of ductwork and other interior elements. The lower level of detail (in which it is possible to know only that a whole floor is complete) may not be adequate for that kind of precision scheduling. Also, it may be that early in the schedule development process, it is not possible to know whether or not multiple sets of formwork will be used. The lower level of detail may be used early on, and higher levels developed as the need arises.

There is no hard and fast rule about how much detail to use. It is a matter of preference among schedulers and project managers. The tendency in the industry is probably to use too little detail, with the result that vital elements of the administrative or procurement process are often overlooked. If high levels of detail are used early in the scheduling process, they can always be simplified later, whereas it is difficult to remember to go back and include something if it was not picked up in the first stages.

System for Description

In developing a list of activities, it is recommended that a consistent system be used to describe them. A system that works well in practice describes each activity in terms of:

1. An action being taken
2. The building element involved in this action
3. A location identifier for the element

Such a system results in the following kinds of activity descriptions:

Place concrete — garage footings
Erect masonry wall — north side
Install tubs — 3rd floor

The following examples show the kind of flexibility a classification system may offer.

Separating Actions

Form foundations — garage area
Place foundations — garage area

In this case, the types of work are kept separate even though the element being worked on and the area are both the same.

Separating the Work Items

Form foundations — garage area
Form columns — garage area

In this case, foundations are kept separate from columns even though both are in the same physical area.

Separating Areas

Form foundations — garage area
Form foundations — dining and kitchen area

In this case, the garage area work is clearly kept separate from the dining/kitchen area, even though both are foundation work.

While this system may seem somewhat rigid, it points up some important distinctions in the activity lists. These issues are often ignored, only to find out later (after several hundred activities have been listed) that the listing, "concrete," has left the user of the schedule with no idea where the concrete is or what phase of the concrete work is being done. It is therefore highly recommended that this system or one similar be used consistently to completely identify the type of work, the object being worked on, and the location of the work.

Tips on Activity List Development
The following guidelines may be helpful in quickly putting together a good activity list.

1. "Brainstorming" can sometimes help. A group of project personnel may meet and, together, record the list on a chalkboard.
2. It is generally better to come up with a large list, and then cull, rather than the other way around.
3. One of the best ways to get a sense of required building activities is to look at the wall sections and read upwards.
4. Procurement and administrative activity lists can often be gleaned from a careful reading of the specifications, taking note of submittal data requirements.
5. The estimate is typically a good source of information about subcontractor work activities, and identifies the responsibilities of each.
6. One should be prepared to go through a series of lists. The first one never has everything right.

Establishing the Sequence of Work

After the activity list for the project has been developed, the scheduler or project manager must decide the order in which the activities will be performed, and communicate that information to those responsible for carrying them out. The construction plan is normally represented by a *Logic Diagram,* or *Network,* which is the basis for the CPM system of scheduling. The diagram is both a tool for making the scheduling decisions and a means of representing the outcome. The diagram will always evolve through a series of versions as the schedule develops and as the project proceeds.

Diagramming Systems
Two methods for representing job logic are in use today. They are commonly known as the *Precedence Diagramming Method* (PDM) and the *Arrow Diagramming Method* (ADM). Either system can be used very effectively on construction projects. There is, however, some debate over the relative merits of these systems, as well as some misunderstanding about the nature of each. Of the two, the arrow system (ADM) is older, and many persons in the industry have used it for most of their careers. It is, however, being rapidly replaced by the precedence system (PDM), which has certain advantages. PDM offers ease of use and understanding, and the capacity to represent a wide variety of job situations. It is for these reasons that PDM is used for all presentations in this book. Most of the techniques for fully representing job logic using PDM are discussed in the text.

Diagramming Formats
Basically, any diagramming technique or format will work in the PDM system so long as it roughly resembles Figure 5.1, in which the activity being represented is shown as a node (typically a rectangle or circle), and the relationship between two or more activities is shown as an arrow between the activities, or nodes. For drawing rough diagrams, a wide range of formats can be used. One option is a work sheet that has pre-drawn nodes; only activity titles and arrows need to be added. The opposite extreme is a blank sheet of drafting paper. Refinement of the diagram as the plan develops typically results in better and more clearly drafted successive versions. Methods of maintaining the logic diagram for clarity and accuracy are covered in Chapter 8.

Development of the Construction Sequence

In order to develop the construction plan, the scheduler or manager need only take each activity in turn and answer the following questions about that activity and related work activities:

1. What other work must be completed before the activity can begin?
2. What other work cannot begin before the activity is completed?
3. What other work can be performed at the same time as the activity and not interfere?

To illustrate this concept, visualize the process of placing the foundation for the sample hotel. The following general activities would be included:

Excavate
Install footings
Erect foundation walls

The first activity, "excavate," is not preceded by any other activity. The next activity is "install footings." None of the other activities can take place at the same time as this one, and it must follow excavation since the footings are below grade. Looking ahead to "erect foundation walls," the scheduler could say that this activity must be preceded by "install footings," since the walls must rest on the footings. The simple relationships between these three activities are illustrated in Figure 5.2.

Expanding the Network

In larger networks, it is possible to become lost in a mass of detail if a proper order is not established for the relationships between activities. It is helpful if the scheduler can view this task in the following terms.

Dealing with Sub-Units of Work

First, it is useful to think of developing the schedule in "chunks," or sub-units. For example, looking at the foundation of the sample hotel project, an

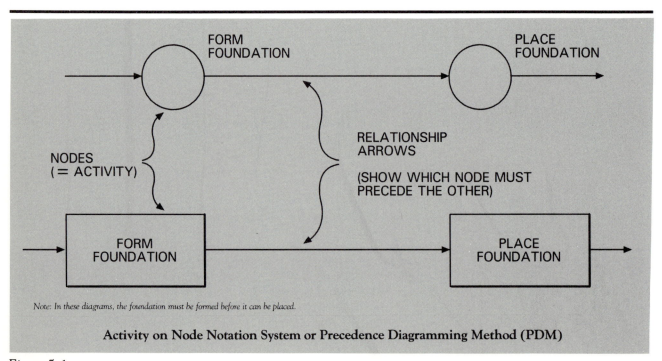

Note: In these diagrams, the foundation must be formed before it can be placed.

Activity on Node Notation System or Precedence Diagramming Method (PDM)

Figure 5.1

experienced construction person knows that the general activities given earlier could be further broken down as follows.

Excavation
— mass excavation
— excavation for trenches
Footings
— column footings
— wall footings
Foundation Walls

This expanded logic is shown in Figure 5.3, the general logic diagram. These sub-units, or "chunks," could be further subdivided. For example, the wall footings might be composed of the following elements:

Form wall footings
Reinforce wall footings
Place wall footings
Strip wall footings

The results of this general-to-detailed expansion process are shown in Figures 5.4 and 5.5. The full expansion process for the entire sample hotel project is shown in Appendix A.

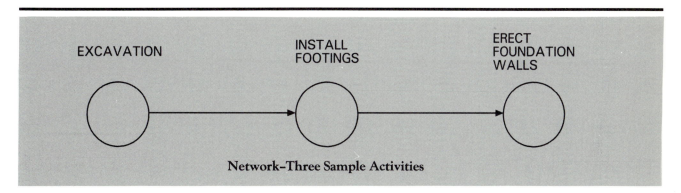

Network–Three Sample Activities

Figure 5.2

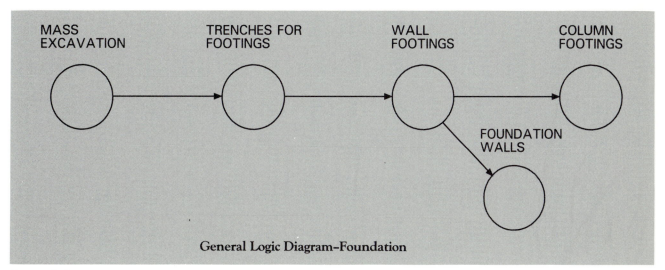

General Logic Diagram–Foundation

Figure 5.3

Priority of Relationships

For a scheduler developing the expanded logic, it is helpful to think in terms of which activity relationships are most important. In addition to the obvious physical relationships between activities, there are others which should logically be considered. Among them are the order of use of various pieces of equipment, weather scheduling problems, and priorities of manpower. These factors involve more complex decision-making techniques. The general principle is that the relationships should be treated in order from the least to the most flexible. The following order of treatment has been proven a most effective approach.

1. Deal with the physical relationships first. These are by far the least flexible. For example, a column must rest on top of a footing if that is what the plan shows; the relationship between these two activities cannot be altered.

2. Deal with the contractual or external relationships next. If there are weather considerations or contractual obligations that must be reflected in the order of activities, these may be altered to some degree.

3. Deal with the managerial and equipment relationships last. Once the previous relationships between activities have been established, the manager can then start to make decisions about the order of equipment and manpower assignments to get the job done. Of all the relationships

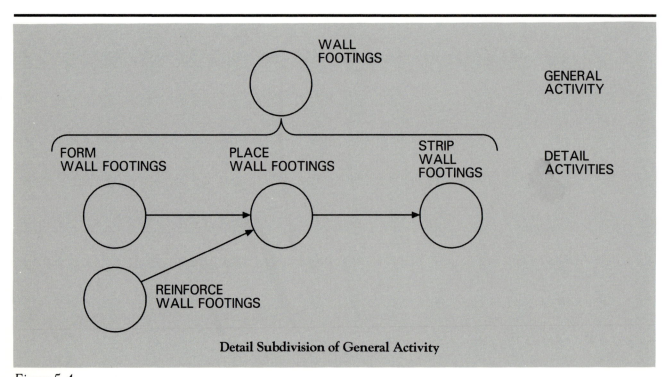

Detail Subdivision of General Activity

Figure 5.4

between activities, it is these last types over which the project manager has the most control.

Complex Relationships between Activities

As noted earlier in this chapter, the relationships between activities depend on what must be finished prior to starting an activity, and what may begin when an activity is complete. This basic idea is at the heart of the CPM concept. One of the advantages of the PDM system of notation is that it permits the construction manager to vary the relationship between activities in order to more fully represent the actual events on a job site.

To consider these advanced forms of activity relationships, a system of definitions must first be established. The standard relationship shown in all the figures of this chapter so far can be defined as a *finish-to-start* relationship. Finish-to start is again illustrated in Figure 5.6, showing the node notation format and a time-scaled bar chart of the relationship between "erect studs" and "hang drywall," which was illustrated earlier in Figure 5.3. Simply put, this standard relationship says that if erecting studs precedes the hanging of drywall, then the studs must be finished before drywall can start.

While this basic relationship applies to many, if not most, activities on a project, there are also many cases where more complex relationships exist, or

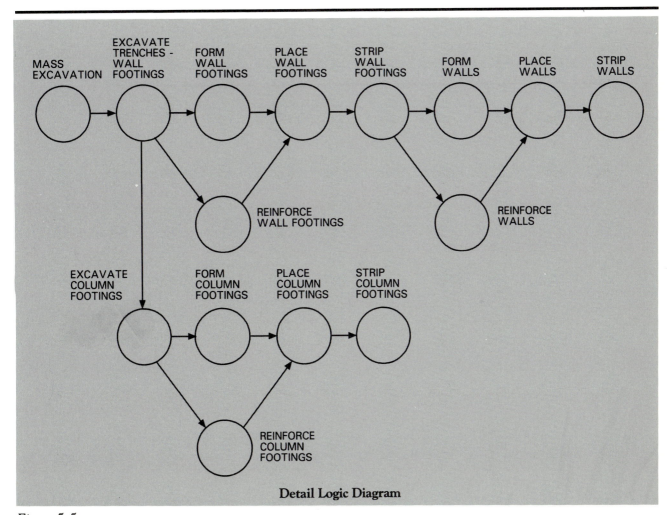

Detail Logic Diagram

Figure 5.5

where the construction manager scheduling the job may want to represent the relationship more realistically. For example, consider the case in which studs and drywall are being installed in a very large building, and there is no point in waiting until all studs are erected before hanging any drywall. It is quite often possible under these circumstances to erect some portion of the studs, say 20% or so, then begin the drywall while stud work proceeds.

This kind of relationship between activities can be shown as *Lag Relationships,* or *overlapping activities.* Instead of showing the relationship between studs and drywall as *finish-to-start* (FS), it can be shown as a *start-to-start* (*SS Lag*). A notation is typically made on the relationship arrow (See Figure 5.7 which also shows the time relationships between the studs and drywall). Another type of lag is the *finish-to-finish* relationship, sometimes called an *F-F lag,* or *FF relationship.* The F-F lag shows a relationship in the finishing times of activities (see Figure 5.8).

Lag relationships can be used anywhere in a network to represent all kinds of situations. It is even possible to have different kinds of lag relationships between the same two activities, if the construction situation requires this arrangement. The most common use of SS and FF relationships is in situations where crews are planned to proceed along a job at different rates of

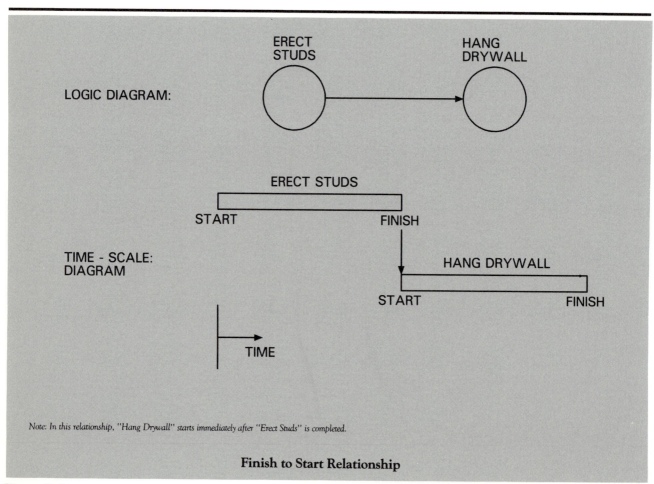

Note: In this relationship, "Hang Drywall" starts immediately after "Erect Studs" is completed.

Finish to Start Relationship

Figure 5.6

production. In this case, the lags are used to establish the correct starting times. The objective is to prevent too great of an initial spacing — which would be inefficient, or overrunning of crews — which would cause interference. Specifically, SS lags are typically used where the succeeding activity is slower than the leading activity, and FF lags are used where the succeeding activity is faster than the leading activity. This point is illustrated in Figure 5.9, where different times are applied to erecting studs and hanging drywall. SS and FF lags are used to show the necessary spacing. In Case 1, it can be seen from the relative times of the two crews that the succeeding drywall crew proceeds more slowly than the stud crew. Consequently, there is no danger of the drywall crew catching up with the stud crew and causing interference. In Case 2, the succeeding drywall crew is faster than the stud crew, and there is a danger of interference if the drywall crew starts too soon. The effect of the FF lag is to delay the start of the drywall crew until such time as interference will not occur.

Tips for Establishing Work Sequences

The following tips may be helpful in working up an effective and realistic network.

1. Make absolutely sure that the plan reflects what the field construction managers want to do with the job. If the people who have to carry out the

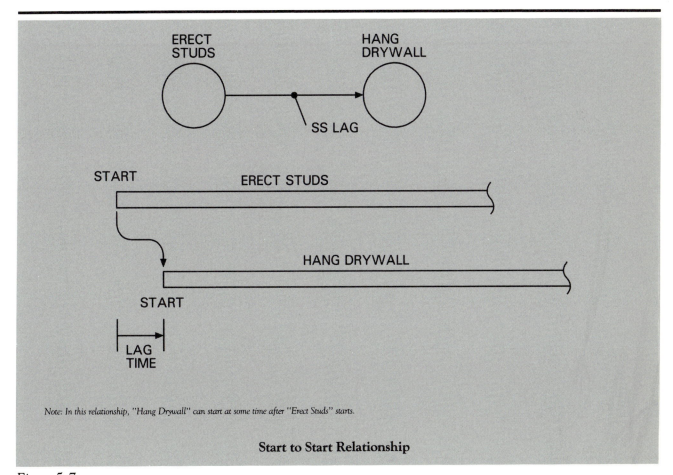

Note: In this relationship, "Hang Drywall" can start at some time after "Erect Studs" starts.

Start to Start Relationship

Figure 5.7

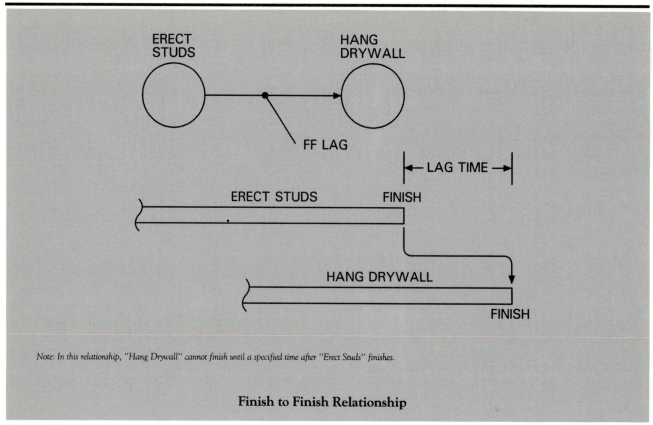

ERECT STUDS

HANG DRYWALL

FF LAG

← LAG TIME →

ERECT STUDS   FINISH

HANG DRYWALL

FINISH

Note: In this relationship, "Hang Drywall" cannot finish until a specified time after "Erect Studs" finishes.

Finish to Finish Relationship

Figure 5.8

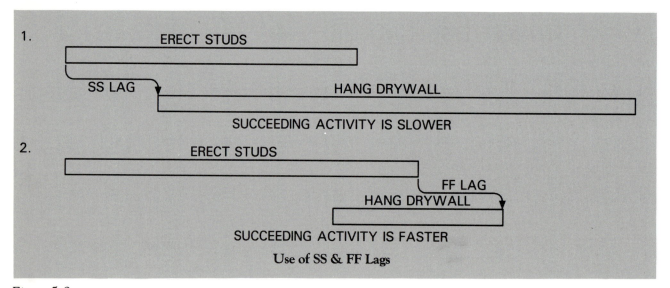

1.

ERECT STUDS

SS LAG

HANG DRYWALL

SUCCEEDING ACTIVITY IS SLOWER

2.

ERECT STUDS

FF LAG

HANG DRYWALL

SUCCEEDING ACTIVITY IS FASTER

Use of SS & FF Lags

Figure 5.9

plan are not committed to it, the chances for success are minimal. Although it is probably best to have the actual managers develop the plan, it is not essential, so long as they are regularly consulted and their ideas are incorporated into the plan.

2. When developing the construction plan, or network, remember that the diagram only represents what is to be done on the job, and that the schedule is not an end in itself. Under no circumstances should scheduling or computer considerations **ever** dictate construction operations.

3. Remember that developing a construction plan is a creative process, and that the first pass is never the final solution. Be prepared to erase, re-draw, and use lots of paper if necessary.

4. There are no hard and fast rules about how the network should be drawn. Some schedulers claim that an arrow should never be drawn in a backwards direction, for example, a statement which is not really valid. What counts is that the diagram is understandable, and while neat drawings are more legible than sloppy ones, anything that works is acceptable practice.

6

Scheduling the Project

6

Scheduling the Project

In the previous chapter, guidelines are proposed for initial job planning. Included are the processes of breaking the project down into workable sub-parts, or activities, and then determining the sequence of work that will be used to accomplish those activities. Once these two tasks are complete, the project manager has created some "tools" which can be of help in finishing the job. This organized approach must be carried further, however, since no time information has yet been developed. This chapter addresses time issues, specifically:

1. estimating the times of individual activities, and
2. calculating the length of the job as a whole.

Techniques are described for accomplishing these two tasks, and for relating work times to calendar times.

Estimating Durations

There are a number of specific methods which can be used to arrive at reasonably accurate estimates of the time required to perform a given activity or piece of work on a job. Before getting into the details of these methods, however, some general rules should be noted. These rules can help to arrive at realistic times and avoid serious errors, and should be observed by anyone developing a construction schedule.

1. Assume Each Activity Will Be Done Normally

First, the time estimates should initially assume a normal or ideal set of working conditions. The reason for using this assumption can be seen in Figure 6.1, which relates efficiency, or unit cost, to activity time. For most activities, there is a most efficient rate of production which results in the lowest possible unit cost. Because all jobs vary somewhat, this rate of production is not a precise figure. Instead, it represents a range of production rates, which, experience has shown, result in the lowest unit cost. Figure 6.1 also shows that if the production rate is significantly higher or lower, the unit cost changes accordingly.

In the best of all possible worlds, a construction manager would expect all the activities on a job to be carried out at a rate close to the ideal. In the real world, the ideal rate is actually possible for most activities. However, those that do not proceed on schedule may be critical to the job, or hold up the progress of other activities in some way. The best procedure, therefore, is to initially plan all activities in terms of ideal time, and then change only those that must be changed for valid reasons, such as overall time.

2. Evaluate Each Activity Independently

In addition to initially assuming an ideal time for each activity, the scheduler should compute individual activity times as if no other work existed. Clearly, this is not realistic in the long run; but practice has shown that much of the work of a project does in fact proceed without being affected very much by other work. It is also true that if a person drawing up a schedule tries initially to consider every constraint affecting an activity, the number of variables may rapidly become overwhelming. Further, many of the factors that will affect a given activity cannot be known until the overall project time has been determined. Thus, it is better to plan each activity independently, and then account for constraints as necessary and as they become known in the scheduling process.

3. Use Consistent Time Units

Throughout this book, the time units used to describe activity and project durations are *workdays*. These workdays are converted to calendar days through the calendar definition process, covered later in this chapter. Days are by far the most common unit of time measure in the construction industry, although hours or weeks can be used appropriately in many circumstances.

Regardless of the time unit used, it is important to be consistent in order to prevent confusion and misunderstanding over the scheduled times for various parts of the job. Calculated times for work activities are usually in workdays; the quoted times for delivery of materials are often in calendar days. The person preparing the schedule must be certain that one is converted, if necessary, to be consistent with the other.

4. Keep Good Records as the Schedule is Developed

It is often helpful during the schedule development process to be able to refer

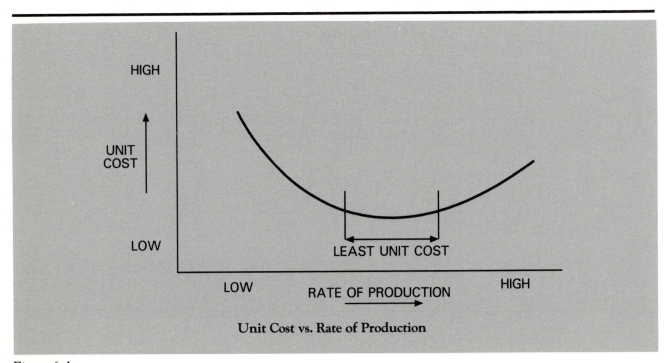

Unit Cost vs. Rate of Production

Figure 6.1

70

to previous assumptions, calculations, and "trial balloons." For example, in trying to decide how much to speed up various activities, it is helpful if one knows what is the assumed normal rate. With this information, it is possible to gauge the effect of one activity's acceleration on the various other activities, so as to decide on the most efficient mix. Toward this end, most good schedulers maintain analysis and record sheets on each activity as the schedule develops. A sample sheet used for this purpose is shown in Figure 6.2.

Typically, any given activity may have several sheets worked out over the course of the job. Dating and keeping each subsequent sheet in order is important. The sheets might be kept in a loose-leaf notebook, catalogued by a classification system, such as the CSI MASTERFORMAT. An organized approach makes referring back to previous data much easier.

Activity Analysis Sheet

CSI Number: _____
Estimate Page Number: _____
Date: _____
Revision Number: _____

Activity Definition:
Action: _____
Object: _____
Location: _____
General Description: _____
_____

Codes Involved:
Cost Code: _____
Schedule Code: _____
Limiting Factors: _____
_____

Total Units of Work: _____
Productivity: MH/Unit: _____   EH/Unit: _____
Calculations:

$$\frac{MH}{Units} \times \underline{\quad} \text{ Total Units} = \underline{\quad} \text{ Total MH}$$

$$\frac{\text{Total MH}}{\text{MH/Crew Day}} = \underline{\quad} \text{ Total Days}$$

Equipment Used: Type: _____   Qty: _____
                    _____   _____
                    _____   _____

Crew Description: _____
_____
_____
_____

Comments and Assumptions: _____
_____
_____
_____
_____
_____

Figure 6.2

Actual Calculation of Activity Durations

There are several methods for determining an accurate activity duration. Each of the methods is applicable in appropriate situations. The emphasis of this book is, however, on *man-hour productivity, or daily production rate-based* methods. These methods have proven to be the most flexible and have the additional advantage that they are based on readily available data from a variety of sources. Possible sources include a company's own historical data, or published information such as *Means Man-Hour Standards.*

Our sample hotel project is used to illustrate the productivity-based techniques and methods. The calculations examples are from the network for the foundation of the sample hotel (developed in the last chapter). An activity list, logic diagram, and summary of all activity durations for the foundation are shown in Figures 6.3 through 6.5.

Man-hour Productivity Method   Traditionally, cost and productivity data within the construction industry have been collected and recorded using the basic *dollars per unit of work in place* (for example, $/S.F., $/C.Y., etc.) method. However, there has been a growing trend toward the use of man-hours per unit of work placed (for example, M.H./S.F., M.H./C.Y., etc.) as a basis for cost estimating and work planning. There are several reasons for this change, including the fact that payroll systems must collect data on a man-hour basis anyway, and inflation has made the traditional use of dollars per unit less reliable and subject to frequent change. The *man-hours per unit measure approach* has proven in practice to be easier to use and more accurate — in both estimating and work planning.

Basic Calculations: The use of man-hour productivity data in estimating activity duration is based on the following formulas:

1. Total man-hours required for an activity =
   Man-hours/unit × units of work for the activity;
   which can be stated mathematically as:
   Total M.H. = M.H./Unit × No. of Units
2. Total days required to finish an activity =
   Total man-hours/man-hours worked per day;
   which can be stated mathematically as:

$$\text{Total Days} = \frac{\text{Total Man-hours}}{\text{Man-hours/day}}$$

For the standard eight-hour workday, the above formula can be restated as:

$$\text{Total Days} = \frac{\text{Total Man-hours}}{\text{Crew Size} \times 8 \text{ hrs/day}}$$

Activity List		
Sample Hotel Foundation		
Mass Excavation	Strip Wall Footings	Form Column Footings
Excavate Trench—Wall Footings	Form Basement Walls	Reinforce Column Footings
Excavate Trench—Column Footings	Reinforce Basement Walls	Place Column Footings
Form Wall Footings	Place Basement Walls	Strip Column Footings
Reinforce Wall Footings	Strip Basement Walls	Cleanup and Complete Foundation
Place Wall Footings		

Figure 6.3

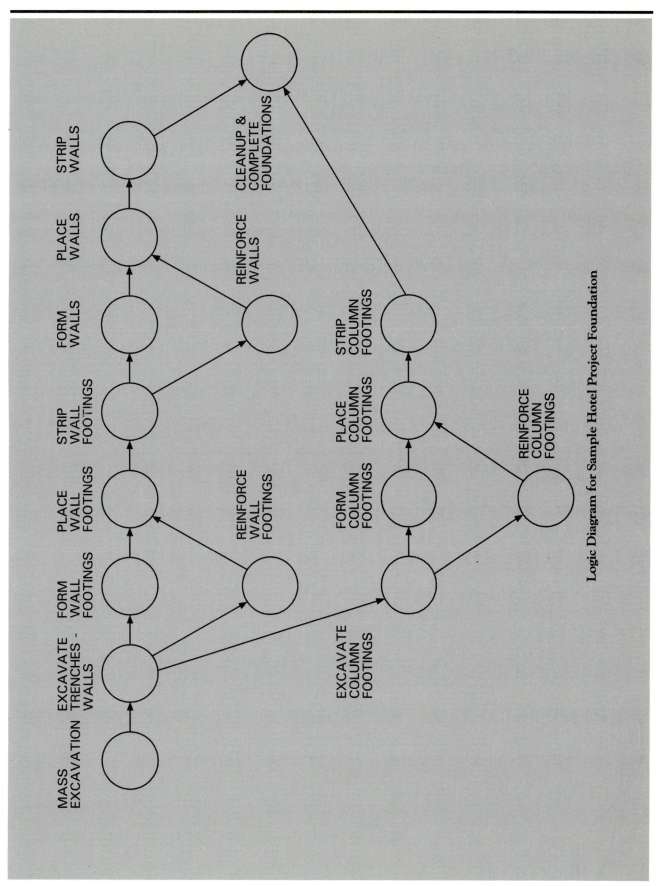

Logic Diagram for Sample Hotel Project Foundation

Figure 6.4

To show how these formulas work, we will use the wall footing activity of the sample foundation network. The total number of square feet of contact area in the wall footings is 1962 (Figure 6.5), a figure that would typically be obtained from the company's estimate. Looking at *Means Man-Hour Standards*, line number 031-158-0150 (Figure 6.6), it can be seen that the man-hour productivity for continuous footings of the type called for in the hotel is .066 man-hours per SFCA installed. Applying these numbers to the basic formulas, we obtain the following results.

Total M.H. = 1962 SFCA × .066 M.H./SFCA = 129.5 Total M.H.
Total days = 129.5/(4 CARPS × 8 hrs) = 4.04 days.

This calculated time of 4.04 days can then be used in work planning, and in the later calculation of overall project time.

Daily Production Rate Method In addition to keeping production information in a man-hour/unit format, many companies have traditionally kept the information in the form of a daily rate of production for a given crew. Publications such as *Means Building Construction Cost Data* also provide this kind of information. The daily crew productivity method is somewhat simpler than using the man-hour productivity method, but it is probably not as flexible, particularly when it comes to varying the crew size by a few

Foundation Network Time Calculations							
	Quan	Unit	Mh/ Unit	Total Mh Req'd	Typical Crew Size	Calc Days Req'd	Rounded Days Req'd
Mass Excavation	9,100	CY	.008	73	1.5	6.1	7
Trench Excavation—Walls	116	CY	.107	12	3	.5	1
Trench Excavation— Columns	42	CY	.107	4	3	.2	1
Formwork: Wall Footings	1,962	SFCA	.050	98	4	3.1	4
Column Footings	1,578	SFCA	.058	92	4	2.9	3
Walls	11,172	SFCA	.092	1,028	12	10.7	11
Concrete: Wall Footings	31	CY	.640	20	8	.3	1
Column Footings	40	CY	1.280	51	8	.8	1
Walls	132	CY	.750	99	8	1.5	2
Reinf: Wall Footings	.8	TNS	15.250	12	2	.8	1
Column Footings	1.0	TNS	15.250	15	2	1.0	1
Walls	2.6	TSN	10.670	28	2	1.7	2
Stripping: Wall Footings	1,962	SFCA	.019	37	4	1.2	2
Column Footings	1,578	SFCA	.017	27	4	.8	1
Walls	11,172	SFCA	.030	335	4	10.5	11

Figure 6.5

		Description	CREW	MAKEUP	DAILY OUTPUT	MAN-HOURS	UNIT
		**031 100	Struct C.I.P. Formwork**					
150	4051	Forms in place, floor slab w/19" metal domes, 2 use	C-2	1 Carpenter Foreman (out) 4 Carpenters 1 Building Laborer Power Tools	435	.110	S.F.
	4100	3 use			465	.103	S.F.
	4150	4 use			495	.097	S.F.
	5000	Box out for slab openings, over 16" deep, 1 use			190	.253	SFCA
	5050	2 use			240	.200	SFCA
	5500	Shallow slab box outs, to 10 S.F.			42	1.140	Ea.
	5550	Over 10 S.F. (use perimeter)			400	.120	L.F.
	6000	Bulkhead forms for slab, with keyway, 1 use, 2 piece			500	.096	L.F.
	6100	3 piece			460	.104	L.F.
	6500	Curb forms, wood, 6" to 12" high, on elevated slabs, 1 use	C-1	3 Carpenters 1 Building Laborer Power Tools	180	.178	SFCA
	6550	2 use			205	.156	SFCA
	6600	3 use			220	.145	SFCA
	6650	4 use			225	.142	SFCA
	7000	Edge forms to 6" high, on elevated slab, 4 use			500	.064	L.F.
	7070	7" to 12" high, 1 use			162	.198	SFCA
	7080	2 use			198	.162	SFCA
	7090	3 use			222	.144	SFCA
	7101	4 use			350	.091	SFCA
	7500	Depressed area forms to 12" high, 4 use			300	.107	L.F.
	7550	12" to 24" high, 4 use			175	.183	L.F.
	8000	Perimeter deck and rail for elevated slabs, straight			90	.356	L.F.
	8050	Curved			65	.492	L.F.
	8500	Void forms, round fiber, 3" diameter			450	.071	L.F.
	8550	4" diameter, void			425	.075	L.F.
	8600	6" diameter, void			400	.080	L.F.
	8650	8" diameter, void			375	.085	L.F.
	8700	10" diameter, void			350	.091	L.F.
	8750	12" diameter, void			300	.107	L.F.
154		FORMS IN PLACE, EQUIPMENT FOUNDATIONS 1 use	C-2	1 Carpenter Foreman (out) 4 Carpenters 1 Building Laborer Power Tools	160	.300	SFCA
	0050	2 use			190	.253	SFCA
	0100	3 use			200	.240	SFCA
	0150	4 use			205	.234	SFCA
158		FORMS IN PLACE, FOOTINGS Continuous wall, 1 use	C-1	3 Carpenters 1 Building Laborer Power Tools	375	.085	SFCA
	0050	2 use			440	.073	SFCA
	0100	3 use			470	.068	SFCA
	0150	4 use			485	.066	SFCA
	0500	Dowel supports for footings or beams, 1 use			500	.064	L.F.
	1000	Integral starter wall, to 4" high, 1 use			400	.080	L.F.
	1500	Keyway, 4 uses, tapered wood, 2" x 4"	1 Carp	1 Carpenter	530	.015	L.F.
	1550	2" x 6"			500	.016	L.F.
	2000	Tapered plastic, 2" x 3"			530	.015	L.F.
	2050	2" x 4"			500	.016	L.F.
	2250	For keyway hung from supports, add			150	.053	L.F.
	3004	Pile cap, square or rectangular, 1 use	C-1	3 Carpenters 1 Building Laborer Power Tools	291	.110	SFCA
	3050	2 use			346	.092	SFCA
	3100	3 use			371	.086	SFCA

Figure 6.6

workers either way in order to adjust times. The following example is based on the same sample activity used to illustrate the man-hour productivity method.

Basic Calculations: The use of daily output or production rate data in estimating activity times is based on this simple formula:

Total Days Required for an Activity = Units of Work for the Activity/Daily Output or, stated mathematically:

$$\text{Total Days} = \frac{\text{Total Units}}{\text{Daily Output}}$$

To illustrate this method, the first activity of the foundation is used once again. The total number of square feet of contact area is 1962. Looking at *Building Construction Cost Data* (1988 edition), Section 031-158-0150 (Figure 6.7), it can be seen that the daily output for the type of footing called for is 485 SFCA per day for a C-1 crew (Figure 6.8), which consists of three carpenters and one laborer. With these numbers applied to the basic formula, the following results are obtained:

$$\text{Total days} = \frac{1962\,\text{SFCA}}{485\,\text{SFCA per day}} = 4.04\,\text{days}$$

This time figure is the same as was calculated using the man-hour productivity method, which is to be expected. By examining the relationship between the two methods, one can see why the results are the same. Crew C-1 consists of four individuals, which means that 4×8 man-hours = 32 man-hours/day are invested in accomplishing the work. Dividing 32 man-hours by 485 SFCA installed per day means that the man-hours/SFCA is .066, which coincides with the value used in the man-hour productivity method.

Experience Method While the two calculation methods given above are very reliable, there is also a place for the method based on experience. Experienced field personnel are very often able to accurately estimate the time required for an activity. This method, while not recommended for large and complex activities, is very useful in estimating unusual kinds of work for which published data may not be available. It is good practice in any case to check with the field personnel on a job to see if the calculated times are reasonable. This safeguard prevents gross calculation errors and serves the additional purpose of bringing the job personnel into the planning process.

Adjustment of Calculated Times

After the basic activity times have been determined, it is a good idea to temper these "rational" calculations with judgments based on the conditions of the particular job. These factors may not be reflected in the basic productivity or production rate data. Since construction is a complex, dynamic process, many calculated activity times must be adjusted to reflect these specific circumstances of the individual job. Some examples of these kinds of considerations follow.

1. **Installing continuous wall forms.** On the sample network used, the installation of wall footings is broken down into forming, reinforcing, placing concrete, and stripping. The two formwork activities — forming and stripping — are separate. Looking at the data from *Means Building Construction Cost Data*, section 031-158-0010 (Figure 6.7) refers to "forms in place." When using any data from a reference source, it pays to determine exactly what that data includes by consulting the explanatory sections. In this case, the explanation is provided in the rear of the book, under Circle Reference Number 33 (Figures 6.9 and 6.10). By consulting the reference section, it is determined that the man-hour productivity of .066 MH/SFCA includes erecting, stripping, cleaning, and moving.

031 | Concrete Formwork

031 100 | Struct C.I.P. Formwork

			CREW	DAILY OUTPUT	MAN-HOURS	UNIT	MAT.	LABOR	EQUIP.	TOTAL	TOTAL INCL O&P	
154	0100	3 use	C-2	200	.240	SFCA	.67	4.98	.14	5.79	8.30	154
	0150	4 use	"	205	.234		.57	4.86	.13	5.56	8	
158	0010	FORMS IN PLACE, FOOTINGS Continuous wall, 1 use	C-1	375	.085		.85	1.71	.05	2.61	3.54	158
	0050	2 use		440	.073		.45	1.46	.05	1.96	2.72	
	0100	3 use		470	.068		.34	1.36	.04	1.74	2.44	
	0150	4 use		485	.066		.28	1.32	.04	1.64	2.32	
	0500	Dowel supports for footings or beams, 1 use		500	.064	L.F.	.52	1.28	.04	1.84	2.52	
	1000	Integral starter wall, to 4" high, 1 use		400	.080		.80	1.60	.05	2.45	3.32	
	1500	Keyway, 4 uses, tapered wood, 2" x 4"	1 Carp	530	.015		.07	.32		.39	.55	
	1550	2" x 6"		500	.016		.09	.34		.43	.60	
	2000	Tapered plastic, 2" x 3"		530	.015		.40	.32		.72	.92	
	2050	2" x 4"		500	.016		.50	.34		.84	1.05	
	2250	For keyway hung from supports, add		150	.053		.50	1.13		1.63	2.23	
	2260											
	3000	Pile cap, square or rectangular, 1 use (33)	C-1	290	.110	SFCA	1.32	2.21	.07	3.60	4.81	
	3050	2 use		346	.092		.76	1.85	.06	2.67	3.66	
	3100	3 use		371	.086		.57	1.73	.05	2.35	3.26	
	3150	4 use		383	.084		.43	1.67	.05	2.15	3.02	
	4000	Triangular or hexagonal caps, 1 use		225	.142		1.55	2.85	.09	4.49	6.05	
	4050	2 use		280	.114		.85	2.29	.07	3.21	4.42	
	4100	3 use		305	.105		.65	2.10	.07	2.82	3.92	
	4150	4 use		315	.102		.55	2.04	.06	2.65	3.71	
	5000	Spread footings, 1 use (33)		305	.105		.95	2.10	.07	3.12	4.25	
	5050	2 use		371	.086		.55	1.73	.05	2.33	3.24	
	5100	3 use		401	.080		.42	1.60	.05	2.07	2.89	
	5150	4 use		414	.077		.35	1.55	.05	1.95	2.75	
	6000	Supports for dowels, plinths or templates, 2' x 2'		25	1.280	Ea.	2.70	26	.82	29.52	42	
	6050	4' x 4' footing		22	1.450		5.80	29	.93	35.73	51	
	6100	8' x 8' footing		20	1.600		11.75	32	1.02	44.77	62	
	6150	12' x 12' footing		17	1.880		19	38	1.20	58.20	78	
	7000	Plinths, 1 use		250	.128	SFCA	1.49	2.56	.08	4.13	5.55	
	7100	4 use		270	.119		.41	2.37	.08	2.86	4.06	
162	0010	FORMS IN PLACE, GRADE BEAM 1 use	C-2	530	.091		1.35	1.88	.05	3.28	4.34	162
	0050	2 use		580	.083		.75	1.72	.05	2.52	3.43	
	0100	3 use		600	.080		.55	1.66	.05	2.26	3.13	
	0150	4 use		605	.079		.45	1.65	.04	2.14	3	
166	0010	FORMS IN PLACE, MAT FOUNDATION 1 use		290	.166		1.30	3.44	.09	4.83	6.65	166
	0050	2 use		310	.155		.70	3.21	.09	4	5.65	
	0100	3 use		330	.145		.52	3.02	.08	3.62	5.15	
	0120	4 use		350	.137		.40	2.85	.08	3.33	4.76	
170	0010	FORMS IN PLACE, SLAB ON GRADE										170
	1000	Bulkhead forms with keyway, 1 use, 2 piece	C-1	510	.063	L.F.	.39	1.26	.04	1.69	2.34	
	1050	3 piece (see also edge forms)		400	.080		.49	1.60	.05	2.14	2.98	
	1100	4 piece		350	.091		.65	1.83	.06	2.54	3.51	
	2000	Curb forms, wood, 6" to 12" high, on grade, 1 use		215	.149	SFCA	1.20	2.98	.09	4.27	5.85	
	2050	2 use		250	.128		.65	2.56	.08	3.29	4.62	
	2100	3 use		265	.121		.50	2.42	.08	3	4.23	
	2150	4 use		275	.116		.45	2.33	.07	2.85	4.05	
	3000	Edge forms, to 6" high, 4 use, on grade		600	.053	L.F.	.18	1.07	.03	1.28	1.83	
	3050	7" to 12" high, 4 use, on grade		435	.074	SFCA	.55	1.47	.05	2.07	2.85	
	3500	For depressed slabs, 4 use, to 12" high		300	.107	L.F.	.46	2.14	.07	2.67	3.76	
	3550	To 24" high		175	.183		.58	3.66	.12	4.36	6.20	
	4000	For slab blockouts, 1 use to 12" high		200	.160		.46	3.21	.10	3.77	5.40	
	4050	To 24" high		120	.267		.58	5.35	.17	6.10	8.75	
	5000	Screed, 24 ga. metal key joint										
	5020	Wood, incl. wood stakes, 1" x 3"	C-1	900	.036	L.F.	.26	.71	.02	.99	1.37	
	5050	2" x 4"	"	900	.036	"	.79	.71	.02	1.52	1.95	

For expanded coverage of these items see *Means Concrete Cost Data 1988*

75

Figure 6.7

CREWS

Crew B-88

Crew No.	Bare Costs Hr.	Daily	Incl. Subs O & P Hr.	Daily	Cost Per Man-hour Bare Costs	Incl. O&P
1 Common Laborer	$16.55	$132.40	$24.60	$196.80	$20.70	$30.51
6 Equip. Oper. (med.)	21.40	1027.20	31.50	1512.00		
2 Feller Bunchers, 50 H.P.		537.60		591.35		
1 Log Chipper, 22" Tree		1564.00		1720.00		
2 Log Skidders, 50 H.P.		545.20		599.70		
1 Dozer, 105 H.P.		221.40		243.55		
1 Chainsaw, Gas, 36" Long		35.40		38.95	51.85	57.03
56 M.H., Daily Totals		$4063.20		$4902.75	$72.55	$87.54

Crew B-89

Crew No.	Hr.	Daily	Hr.	Daily	Bare Costs	Incl. O&P
1 Equip. Oper. (light)	$20.20	$161.60	$29.75	$238.00	$18.70	$27.50
1 Truck Driver (light)	17.20	137.60	25.25	202.00		
1 Truck, Stake Body, 3 Ton		158.60		174.45		
1 Concrete Saw		83.20		91.50		
1 Water Tank, 65 Gal		6.00		6.60	15.48	17.03
16 M.H., Daily Totals		$547.00		$712.55	$34.18	$44.53

Crew B-89A

Crew No.	Hr.	Daily	Hr.	Daily	Bare Costs	Incl. O&P
1 Skilled Worker	$21.45	$171.60	$32.00	$256.00	$19.00	$28.30
1 Laborer	16.55	132.40	24.60	196.80		
1 Core Drill (large)		42.60		46.85	2.66	2.92
16 M.H., Daily Totals		$346.60		$499.65	$21.66	$31.22

Crew B-90

Crew No.	Hr.	Daily	Hr.	Daily	Bare Costs	Incl. O&P
1 Labor Foreman (outside)	$18.55	$148.40	$27.55	$220.40	$17.93	$26.50
3 Highway Laborers	16.55	397.20	24.60	590.40		
2 Equip. Oper. (light)	20.20	323.20	29.75	476.00		
2 Truck Drivers (heavy)	17.45	279.20	25.60	409.60		
1 Road Mixer, 310 H.P.		692.60		761.85		
1 Dist. Truck, 2000 Gal.		237.80		261.60	14.53	15.99
64 M.H., Daily Totals		$2078.40		$2719.85	$32.46	$42.49

Crew B-90A

Crew No.	Hr.	Daily	Hr.	Daily	Bare Costs	Incl. O&P
1 Labor Foreman	$18.55	$148.40	$27.55	$220.40	$19.60	$28.96
2 Laborers	16.55	264.80	24.60	393.60		
4 Equip. Oper. (medium)	21.40	684.80	31.50	1008.00		
2 Graders, 30,000 lbs		942.40		1036.65		
1 Roller, Steel Wheel		153.60		168.95		
1 Roller, Pneumatic Wheel		182.20		200.40	22.82	25.10
56 M.H., Daily Totals		$2376.20		$3028.00	$42.42	$54.06

Crew B-90B

Crew No.	Hr.	Daily	Hr.	Daily	Bare Costs	Incl. O&P
1 Labor Foreman	$18.55	$148.40	$27.55	$220.40	$19.30	$28.54
2 Laborers	16.55	264.80	24.60	393.60		
3 Equip. Oper. (medium)	21.40	513.60	31.50	756.00		
1 Roller, Steel Wheel		153.60		168.95		
1 Roller, Pneumatic Wheel		182.20		200.40		
1 Road Mixer, 310 H.P.		692.60		761.85	21.42	23.56
48 M.H., Daily Totals		$1955.20		$2501.20	$40.72	$52.10

Crew B-91

Crew No.	Hr.	Daily	Hr.	Daily	Bare Costs	Incl. O&P
1 Labor Foreman (outside)	$18.55	$148.40	$27.55	$220.40	$19.33	$28.54
2 Highway Laborers	16.55	264.80	24.60	393.60		
4 Equip. Oper. (med.)	21.40	684.80	31.50	1008.00		
1 Truck Driver (heavy)	17.45	139.60	25.60	204.80		
1 Dist. Truck, 3000 Gal.		260.60		286.65		
1 Aggreg. Spreader, S.P.		419.00		460.90		
1 Roller, Pneu. Tire, 12 Ton		182.20		200.40		
1 Roller, Steel, 10 Ton		153.60		168.95	15.86	17.45
64 M.H., Daily Totals		$2253.00		$2943.70	$35.19	$45.99

Crew B-92

Crew No.	Bare Costs Hr.	Daily	Incl. Subs O & P Hr.	Daily	Cost Per Man-hour Bare Costs	Incl. O&P
1 Labor Foreman (outside)	$18.55	$148.40	$27.55	$220.40	$17.05	$25.33
3 Highway Laborers	16.55	397.20	24.60	590.40		
1 Crack Cleaner, 25 H.P.		61.40		67.55		
1 Air Compressor		40.00		44.00		
1 Tar Kettle, T.M.		49.00		53.90		
1 Flatbed Truck, 3 Ton		158.60		174.45	9.65	10.62
32 M.H., Daily Totals		$854.60		$1150.70	$26.70	$35.95

Crew B-93

Crew No.	Hr.	Daily	Hr.	Daily	Bare Costs	Incl. O&P
1 Equip. Oper. (med.)	$21.40	$171.20	$31.50	$252.00	$21.40	$31.50
1 Feller Buncher, 50 H.P.		268.80		295.70	33.60	36.96
8 M.H., Daily Totals		$440.00		$547.70	$55.00	$68.46

Crew C-1

Crew No.	Hr.	Daily	Hr.	Daily	Bare Costs	Incl. O&P
3 Carpenters	$21.20	$508.80	$31.50	$756.00	$20.03	$29.77
1 Building Laborer	16.55	132.40	24.60	196.80		
Power Tools		20.40		22.45	.63	.70
32 M.H., Daily Totals		$661.60		$975.25	$20.66	$30.47

Crew C-1A

Crew No.	Hr.	Daily	Hr.	Daily	Bare Costs	Incl. O&P
1 Carpenter	$21.20	$169.60	$31.50	$252.00	$21.20	$31.50
1 Circular Saw, 7"		6.80		7.50	.85	.93
8 M.H., Daily Totals		$176.40		$259.50	$22.05	$32.43

Crew C-2

Crew No.	Hr.	Daily	Hr.	Daily	Bare Costs	Incl. O&P
1 Carpenter Foreman (out)	$23.20	$185.60	$34.45	$275.60	$20.75	$30.84
4 Carpenters	21.20	678.40	31.50	1008.00		
1 Building Laborer	16.55	132.40	24.60	196.80		
Power Tools		27.20		29.90	.56	.62
48 M.H., Daily Totals		$1023.60		$1510.30	$21.31	$31.46

Crew C-3

Crew No.	Hr.	Daily	Hr.	Daily	Bare Costs	Incl. O&P
1 Rodman Foreman	$24.60	$196.80	$39.15	$313.20	$21.03	$32.76
4 Rodmen (reinf.)	22.60	723.20	36.00	1152.00		
1 Equip. Oper. (light)	20.20	161.60	29.75	238.00		
2 Building Laborers	16.55	264.80	24.60	393.60		
Stressing Equipment		31.95		35.15		
Grouting Equipment		101.20		111.30	2.08	2.28
64 M.H., Daily Totals		$1479.55		$2243.25	$23.11	$35.04

Crew C-4

Crew No.	Hr.	Daily	Hr.	Daily	Bare Costs	Incl. O&P
1 Rodman Foreman	$24.60	$196.80	$39.15	$313.20	$23.10	$36.78
3 Rodmen (reinf.)	22.60	542.40	36.00	864.00		
Stressing Equipment		31.95		35.15	.99	1.09
32 M.H., Daily Totals		$771.15		$1212.35	$24.09	$37.87

Crew C-5

Crew No.	Hr.	Daily	Hr.	Daily	Bare Costs	Incl. O&P
1 Rodman Foreman	$24.60	$196.80	$39.15	$313.20	$22.12	$34.55
4 Rodmen (reinf.)	22.60	723.20	36.00	1152.00		
1 Equip. Oper. (crane)	21.90	175.20	32.25	258.00		
1 Equip. Oper. Oiler	18.00	144.00	26.50	212.00		
1 Hyd. Crane, 25 Ton		439.00		482.90	7.83	8.62
56 M.H., Daily Totals		$1678.20		$2418.10	$29.95	$43.17

Crew C-6

Crew No.	Hr.	Daily	Hr.	Daily	Bare Costs	Incl. O&P
1 Labor Foreman (outside)	$18.55	$148.40	$27.55	$220.40	$17.50	$25.82
4 Building Laborers	16.55	529.60	24.60	787.20		
1 Cement Finisher	20.25	162.00	29.00	232.00		
2 Gas Engine Vibrators		51.60		56.75	1.07	1.18
48 M.H., Daily Totals		$891.60		$1296.35	$18.57	$27.00

Figure 6.8

㉜ Forms for Reinforced Concrete (Div. 031)

Design Economy

Avoid many sizes in proportioning beams and columns.

From story to story avoid changing column dimensions. Gain strength by adding steel or using a richer mix. If a change in size of column is necessary vary one dimension only to minimize form alterations. Keep beams and columns the same width.

From floor to floor in a multi-story building vary beam depth not width as that will leave slab panel form unchanged. It is cheaper to vary the strength of a beam from floor to floor by means of steel area than by 2″ changes in either width or depth.

Cost Factors

Material includes the cost of lumber, cost of rent or metal pans or forms if used, nails, form ties, form oil, bolts and accessories.

Labor includes the cost of carpenters to make up, erect, remove and repair, plus common labor to clean and move. Having carpenters remove forms minimizes repairs.

Improper alignment and condition of forms will increase finishing cost. When forms are heavily oiled, concrete surfaces must be neutralized before finishing. Special curing compounds will cause spillages to spall off in first frost. Gang forming methods will reduce costs on large projects.

Materials Used

Boards are seldom used unless their architectural finish is required. Generally, steel, fiberglass and plywood are used for contact surfaces. Labor on plywood is 10% less than with boards. The plywood is backed up with 2 x 4's at 12″ to 32″ O.C. Walers are generally 2 - 2 x 4's. Column forms are held together with steel yokes or bands. Shoring is with adjustable shoring or scaffolding for high ceilings.

Reuse

Floor and column forms can be reused four or possibly five times without excessive repair. Remember to allow for 10% waste on each reuse.

When modular sized wall forms are made, up to twenty uses can be expected with exterior plyform.

When forms are reused, the cost to erect, strip, clean and move will not be affected. 10% replacement of lumber should be included and about one hour of carpenter time for repairs on each reuse per 100 S.F.

The reuse cost for certain accessory items normally rented on a monthly basis will be lower than the cost for the first use.

After fifth use, new material required plus time needed for repair prevent form cost from dropping further and it may go up. Much depends on care in stripping, the number of special bays, changes in beam or column sizes and other factors.

1. Costs for multiple use of formwork may be developed as follows:

2 Uses	3 Uses	4 Uses
$\dfrac{\text{1st Use + Reuse}}{2} = \text{avg. cost/2 uses}$	$\dfrac{\text{1st Use + 2 Reuse}}{3} = \text{avg. cost/3 uses}$	$\dfrac{\text{1st Use + 3 Reuse}}{4} = \text{avg. cost/4 uses}$

㉝ Forms In Place (Div. 031)

This section assumes that all cuts are made with power saws, that adjustable shores are employed and that maximum use is made of commercial form ties and accessories. Bare costs are used in the table below.

BEAM AND GIRDER, INTERIOR, 12″ Wide (Line 138-2000)	First Use			Reuse		
	Quantities	Material	Installation	Quantities	Material	Installation
5/8″ exterior plyform at $650 per M.S.F.	115 S.F.	$ 74.75		11.5 S.F.	$ 7.50	
Lumber at $340 per M.B.F.	200 B.F.	68.00		20.0 B.F.	6.80	
Accessories, incl. adjustable shores	Allow	20.65		Allow	20.65	
Make up, crew C-2 at $21.31 per man-hour	6.4 M.H.		$136.40	1.0 M.H.		$ 21.30
Erect and strip	8.3 M.H.		176.90	8.3 M.H.		176.90
Clean and move	1.3 M.H.		27.70	1.3 M.H.		27.70
Total per 100 S.F.C.A.	16.0 M.H.	$163.40	$341.00	10.6 M.H.	$34.95	$225.90

For structural steel frame with beams encased, subtract 1.2 man-hours, and 50 M.B.F. lumber or about $40 per 100 S.F.C.A. for the first use and $26 for each reuse.

BOX CULVERT, 5' to 8' Square or Rectangular (Line 146-0010)	First Use			Reuse		
	Quantities	Material	Installation	Quantities	Material	Installation
3/4″ exterior plyform at $705 per M.S.F.	110 S.F.	$ 77.55		11.0 S.F.	$ 7.75	
Lumber at $340 per M.B.F.	170 B.F.	57.80		17.0 B.F.	5.80	
Accessories	Allow	16.25		Allow	16.25	
Build in place, crew C-1 at $20.66 per man-hour	14.5 M.H.		$299.55	14.5 M.H.		$299.55
Strip and salvage	4.3 M.H.		88.85	4.3 M.H.		88.85
Total per 100 S.F.C.A.	18.8 M.H.	$151.60	$388.40	18.8 M.H.	$29.80	$388.40

Figure 6.9

(33) Forms in Place (cont.)

COLUMNS, 24" x 24" (Line 142-6500)	First Use			Reuse		
	Quantities	Material	Installation	Quantities	Material	Installation
5/8" exterior plyform @ $650 per M.S.F.	120 S.F.	$ 78.00		12.0 S.F.	$ 7.80	
Lumber @ $340 per M.B.F.	125 B.F.	42.50		12.5 B.F.	4.25	
Clamps, chamfer strips and accessories	Allow	20.80		Allow	4.60	
Make up, crew C-1 at $20.66 per man-hour	5.8 M.H.		$119.85	1.0 M.H.		$ 20.65
Erect and strip	9.8 M.H.		202.45	9.8 M.H.		202.45
Clean and move	1.2 M.H.		24.80	1.2 M.H.		24.80
Total per 100 S.F.C.A.	16.8 M.H.	$141.30	$347.10	12.0 M.H.	$16.65	$247.90

FLAT SLAB WITH DROP PANELS (Line 150-2000)	First Use			Reuse		
	Quantities	Material	Installation	Quantities	Material	Installation
5/8" exterior plyform @ $650 per M.S.F.	115 S.F.	$ 74.75		11.5 S.F.	$ 7.50	
Lumber @ $340 per M.B.F.	210 B.F.	71.40		21 B.F.	7.15	
Accessories, incl. adjustable shores	Allow	22.40		Allow	22.40	
Make up, crew C-2 at $21.31 per man-hour	3.5 M.H.		$ 74.60	1.0 M.H.		$ 21.30
Erect and strip	6.0 M.H.		127.85	6.0 M.H.		127.85
Clean and move	1.2 M.H.		25.55	1.2 M.H.		25.55
Total per 100 S.F.C.A.	10.7 M.H.	$168.55	$228.00	8.2 M.H.	$37.05	$174.70

Drop panels included but column caps figure with columns.

FOOTINGS, SPREAD Line (158-5000)	First Use			Reuse		
	Quantities	Material	Installation	Quantities	Material	Installation
Lumber @ $340 per M.B.F.	260 B.F.	$ 88.40		26 B.F.	$ 8.85	
Accessories	Allow	6.50		Allow	6.50	
Make up, crew C-1 at $20.66 per man-hour	4.7 M.H.		$ 97.10	1.0 M.H.		$ 20.65
Erect and strip	4.2 M.H.		86.75	4.2 M.H.		86.75
Clean and move	1.6 M.H.		33.05	1.6 M.H.		33.05
Total per 100 S.F.C.A.	10.5 M.H.	$ 94.90	$216.90	6.8 M.H.	$15.35	$140.45

FOUNDATION WALL, 8' High (Line 182-2000)	First Use			Reuse		
	Quantities	Material	Installation	Quantities	Material	Installation
5/8" exterior plyform @ $650 per M.S.F.	110 S.F.	$ 71.50		11.0 S.F.	$ 7.15	
Lumber @ $340 per M.B.F.	140 B.F.	47.60		14.0 B.F.	4.75	
Accessories	Allow	16.40		Allow	16.40	
Make up, crew C-2 at $21.31 per man-hour	5.0 M.H.		$106.55	1.0 M.H.		$ 21.30
Erect and strip	6.5 M.H.		138.50	6.5 M.H.		138.50
Clean and move	1.5 M.H.		31.95	1.5 M.H.		31.95
Total per 100 S.F.C.A.	13.0 M.H.	$135.50	$277.00	9.0 M.H.	$28.30	$191.75

PILE CAPS, Square or Rectangular (Line 158-3000)	First Use			Reuse		
	Quantities	Material	Installation	Quantities	Material	Installation
5/8" exterior plyform @ $650 per M.S.F.	110 S.F.	$ 71.50		11.0 S.F.	$ 7.15	
Lumber @ $340 per M.B.F.	160 B.F.	54.40		16.0 B.F.	5.45	
Accessories	Allow	6.50		Allow	6.50	
Make up, crew C-1 at $20.66 per man-hour	4.5 M.H.		$ 92.95	1.0 M.H.		$ 20.65
Erect and strip	5.0 M.H.		103.30	5.0 M.H.		103.30
Clean and move	1.5 M.H.		31.00	1.5 M.H.		31.00
Total per 100 S.F.C.A.	11.0 M.H.	$132.40	$227.25	7.5 M.H.	$19.10	$154.95

STAIRS, Average Run (Inclined Length x Width) (Line 174-0010)	First Use			Reuse		
	Quantities	Material	Installation	Quantities	Material	Installation
5/8" exterior plyform @ $650 per M.S.F.	110 S.F.	$ 71.50		11.0 S.F.	$ 7.15	
Lumber @ $340 per M.B.F.	425 B.F.	144.50		42.5 B.F.	14.45	
Accessories	Allow	16.50		Allow	16.50	
Build in place, crew C-2 at $21.31 per man-hour	25.0 M.H.		$532.75	25.0 M.H.		$532.75
Strip and salvage	4.0 M.H.		85.25	4.0 M.H.		85.25
Total per 100 S.F.	29.0 M.H.	$232.50	$618.00	29.0 M.H.	$38.10	$618.00

Figure 6.10

Since the activities in the sample network are broken down into forming (erecting) and stripping, the .066 MH/SFCA must be divided between these two activities. A good rule of thumb for time apportionment of forming and stripping is 75/25; therefore the productivity time for forming can be calculated as: .066 MH/SFCA × 75% = .049 MH/SFCA. The productivity for stripping can be calculated as: .066 × 25% = .017 M.H./SFCA.

Further examples of time adjustments using the network for the interior of a typical floor in the sample hotel are described below.

2. **Erecting hollow metal door frames.** In some cases, it is not worth investing the time required to precision estimate productivity time. For example, there are two widths of hollow metal door frames on the project, 2'8" and 3'0". The productivities for these two door sizes are so similiar that they are listed under one line number in *Means Man-Hour Standards* (2nd edition): 081-118-0100 (See Figure 6.11).

3. **Spraying the acoustical ceiling.** *Means Man-Hour Standards* lists minimum and maximum values of .046 M.H./S.F. and .120 M.H./S.F., respectively. This is a variation of almost 100%, but the lower figure is based on spraying onto bare concrete, a fairly straightforward and, therefore, quick operation. (See Figure 6.12)

While the examples given here are relatively simple, the point remains valid for all data used as it is input to the scheduling process. Also, in the event the data is altered to fit circumstances, it is better to err on the side of conservatism rather than over-optimism. There is a definite practical and psychological value in building small amounts of "cushion" time into schedules.

Rounding Up all Times   The times obtained in the sample calculations for footings (using the man-hour productivity method) should, in practice, be rounded up to the next higher number of days. This would mean using a scheduled time of five days per floor in the case of the wall footings. In actual practice, it would probably take a little more than 4.04 days to set up, initially get up to speed, do the actual work, and then take down the equipment, etc. Also, using five days tends to build a little bit of a cushion into the overall schedule, time that can help to accommodate the inevitable problems in getting crews going.

More Than One Type of Work in the Activity   Not all reference material contains man-hour productivity data in exactly the same format as the work is broken down in the scheduler's activity list. An example, bathroom accessories installation times from the sample interior network, is shown in Figure 6.13, a page from *Means Man-Hour Standards*, 2nd edition. This information is organized by individual item; there is no entry for accessories for an entire bathroom. It is, however, logical to assign the work of installing all the accessories for a single bathroom or for a group of bathrooms to one carpenter. The total number of man-hours for the task must, therefore, be built up by adding together all the pertinent times, as shown below.

Curtain Rod	.615
Towel Bar — 3 ea × .381 M.H.	1.143
Mirror	1.330
Robe Hooks — 3 ea × .222 M.H.	.666
Toilet Tissue Dispenser	.333
Towel Shelf	.400
Total man-hours per bathroom	4.487

081 | Metal Doors and Frames

081 100	Steel Doors And Frames	CREW	MAKEUP		DAILY OUTPUT	MAN-HOURS	UNIT
110 0521	Fire door, flush, "B" label, 90 min., comp., 20 ga., 2' x 6'-8"	F-2	2 Carpenters Power Tools		18	.889	Ea.
0540	2'-6" x 6'-8"				17	.941	Ea.
0560	3'-0" x 6'-8"				16	1.000	Ea.
0580	3'-0" x 7'-0"				16	1.000	Ea.
0640	Flush, "A" label 3 hour, composite, 18 ga., 3'-0" x 6'-8"				15	1.070	Ea.
0660	2'-6" x 7'-0"				16	1.000	Ea.
0680	3'-0" x 7'-0"				15	1.070	Ea.
0700	4'-0" x 7'-0"				14	1.140	Ea.
114	RESIDENTIAL DOOR Steel, 24 ga., embos., full, 2'-8" x 6'-8"				16	1.000	Ea.
0040	3'-0" x 6'-8"				15	1.070	Ea.
0060	3'-0" x 7'-0"				15	1.070	Ea.
0220	Half glass, 2'-8" x 6'-8"				17	.941	Ea.
0240	3'-0" x 6'-8"				16	1.000	Ea.
0260	3'-0" x 7'-0"				16	1.000	Ea.
0720	Raised plastic face, full panel, 2'-8" x 6'-8"				16	1.000	Ea.
0740	3'-0" x 6'-8"				15	1.070	Ea.
0760	3'-0" x 7'-0"				15	1.070	Ea.
0820	Half glass, 2'-8" x 6'-8"				17	.941	Ea.
0840	3'-0" x 6'-8"				16	1.000	Ea.
0860	3'-0" x 7'-0"				16	1.000	Ea.
1320	Flush face, full panel, 2'-6" x 6'-8"				16	1.000	Ea.
1340	3'-0" x 6'-8"				15	1.070	Ea.
1360	3'-0" x 7'-0"				15	1.070	Ea.
1420	Half glass, 2'-8" x 6'-8"				17	.941	Ea.
1440	3'-0" x 6'-8"				16	1.000	Ea.
1460	3'-0" x 7'-0"				16	1.000	Ea.
2300	Interior, residential, closet, bi-fold, 6'-8" x 2'-0" wide				16	1.000	Ea.
2330	3'-0" wide				16	1.000	Ea.
2360	4'-0" wide				15	1.070	Ea.
2400	5'-0" wide				14	1.140	Ea.
2420	6'-0" wide	↓	↓		13	1.230	Ea.
118	STEEL FRAMES, KNOCK DOWN 18 ga., up to 5-3/4" deep						
0025	6'-8" high, 3'-0" wide, single	F-2	2 Carpenters Power Tools		16	1.000	Ea.
0040	6'-0" wide, double				14	1.140	Ea.
0100	7'-0" high, 3'-0" wide, single				16	1.000	Ea.
0140	6'-0" wide, double				14	1.140	Ea.
2800	18 ga. drywall, up to 4-7/8" deep, 7'-0" high, 3'-0" wide, single				16	1.000	Ea.
2840	6'-0" wide, double				14	1.140	Ea.
3600	16 ga., up to 5-3/4" deep, 7'-0" high, 4'-0" wide, single				15	1.070	Ea.
3640	8'-0" wide, double				12	1.330	Ea.
3700	8'-0" high, 4'-0" wide, single				15	1.070	Ea.
3740	8'-0" wide, double				12	1.330	Ea.
4000	6-3/4" deep, 7'-0" high, 4'-0" wide, single				15	1.070	Ea.
4040	8'-0" wide, double				12	1.330	Ea.
4100	8'-0" high, 4'-0" wide, single				15	1.070	Ea.
4140	8'-0" wide, double				12	1.330	Ea.
4400	8-3/4" deep, 7'-0" high, 4'-0" wide, single				15	1.070	Ea.
4440	8'-0" wide, double				12	1.330	Ea.
4500	8'-0" high, 4'-0" wide, single				15	1.070	Ea.
4540	8'-0" wide, double				12	1.330	Ea.
4800	16 ga. drywall, up to 3-7/8" deep, 7'-0" high, 3'-0" wide, single				16	1.000	Ea.
4840	6'-0" wide, double				14	1.140	Ea.
5400	16 ga. "B" label, up to 5-3/4" deep, 7'-0" high, 4'-0" wide, single				15	1.070	Ea.
5440	8'-0" wide, double				12	1.330	Ea.
5800	6-3/4" deep, 7'-0" high, 4'-0" wide, single				15	1.070	Ea.
5840	8'-0" wide, double	↓	↓		12	1.330	Ea.

Figure 6.11

072 200 | Roof & Deck Insulation

			CREW	MAKEUP	DAILY OUTPUT	MAN-HOURS	UNIT
203	1801	Roof deck insulation, phenolic foam, 4' x 8' sheets					
	1810	1-3/16" thick, R10	1 Rofc	1 Roofer, Composition	1,000	.008	S.F.
	1820	1-1/2" thick, R12.5			1,000	.008	S.F.
	1830	1-3/4" thick, R14.6			1,000	.008	S.F.
	1840	2" thick, R16.7			800	.010	S.F.
	1850	2-1/2" thick, R20			800	.010	S.F.
	1860	3" thick, R25	↓	↓	800	.010	S.F.
	1900	Polystyrene					
	1910	Extruded, 2.3#/C.F., 1" thick, R5.26	1 Rofc	1 Roofer, Composition	1,500	.005	S.F.
	1920	2" thick, R10			1,250	.006	S.F.
	1930	3" thick, R15			1,000	.008	S.F.
	2010	Expanded bead board, 1" thick, R3.57			1,500	.005	S.F.
	2100	2" thick, R7.14	↓	↓	1,250	.006	S.F.
	2200	Urethane, felt both sides					
	2210	1" thick, R6.7	1 Rofc	1 Roofer, Composition	1,000	.008	S.F.
	2220	1-1/2" thick, R11.11			1,000	.008	S.F.
	2230	2" thick, R14.3			800	.010	S.F.
	2240	2-1/2" thick, R20			800	.010	S.F.
	2250	3" thick, R25	↓	↓	800	.010	S.F.
	2300	Urethane and gypsum board composite					
	2310	1-5/8" thick, R7.7	1 Rofc	1 Roofer, Composition	1,000	.008	S.F.
	2320	2" thick, R10			800	.010	S.F.
	2330	2-1/2" thick, R14.3			800	.010	S.F.
	2340	3" thick, R18.2	↓	↓	800	.010	S.F.

072 400 | Exterior Insulation

			CREW	MAKEUP	DAILY OUTPUT	MAN-HOURS	UNIT
402	0020	INTEGRATED SIDING Fabric reinforced synthetic exterior finish, on 1" polystyrene insulation board					
	0100	Minimum	J-1	3 Plasterers 2 Plasterer Helpers 1 Mixing Machine, 6 C.F.	380	.105	S.F.
	0200	Maximum	"	"	270	.148	S.F.
	0300	For insulation, 2" polystyrene, add	1 Plas	1 Plasterer	725	.011	S.F.

072 550 | Cement Fireproofing

			CREW	MAKEUP	DAILY OUTPUT	MAN-HOURS	UNIT
554	0050	SPRAYED Mineral fiber or cementitious for fireproofing, not incl. tamping or canvas protection					
	0100	1" thick, on flat plate steel	G-2	1 Plasterer 1 Plasterer Helper 1 Building Laborer Grouting Equipment	3,000	.008	S.F.
	0200	Flat decking			2,400	.010	S.F.
	0400	Beams			1,500	.016	S.F.
	0500	Corrugated or fluted decks			1,250	.019	S.F.
	0700	Columns, 1-1/8" thick			1,100	.022	S.F.
	0800	2-3/16" thick			700	.034	S.F.
	0900	For canvas protection, add			5,000	.005	S.F.
	1000	Acoustical sprayed, 1" thick, finished, straight work, minimum			520	.046	S.F.
	1100	Maximum			200	.120	S.F.
	1300	Difficult access, minimum			225	.107	S.F.
	1400	Maximum	↓	↓	130	.185	S.F.
	1500	Intumescent epoxy fireproofing on wire mesh, 3/16" thick					
	1550	1 hour rating, exterior use	G-2	1 Plasterer 1 Plasterer Helper 1 Building Laborer Grouting Equipment	136	.176	S.F.
	1600	Magnesium oxychloride, 35# to 40# density, 1/4" thick	↓	↓	3,000	.008	S.F.
	1650	1/2" thick	↓	↓	2,000	.012	S.F.

Figure 6.12

108 200	Bath Accessories	CREW	MAKEUP	DAILY OUTPUT	MAN-HOURS	UNIT
204	**BATHROOM ACCESSORIES**					
0200	Curtain rod, stainless steel, 5' long, 1" diameter	1 Carp	1 Carpenter	13	.615	Ea.
0300	1-1/4" diameter	"	"	13	.615	Ea.
0500	Dispenser units, combined soap & towel dispensers,					
0510	mirror and shelf, flush mounted	1 Carp	1 Carpenter	10	.800	Ea.
0600	Towel dispenser and waste receptacle,					
0610	flush mounted	1 Carp	1 Carpenter	10	.800	Ea.
0800	Grab bar, straight, 1" diameter, stainless steel, 12" long			24	.333	Ea.
0900	18" long			23	.348	Ea.
1000	24" long			22	.364	Ea.
1100	36" long			20	.400	Ea.
1200	1-1/2" diameter, 18" long			23	.348	Ea.
1300	36" long			20	.400	Ea.
1500	Tub bar, 1" diameter, horizontal			14	.571	Ea.
1600	Plus vertical arm			12	.667	Ea.
1900	End tub bar, 1" diameter, 90° angle			12	.667	Ea.
2300	Hand dryer, surface mounted, electric, 110 volt			4	2.000	Ea.
2400	220 volt			4	2.000	Ea.
2600	Hat and coat strip, stainless steel, 4 hook, 36" long			24	.333	Ea.
2700	6 hook, 60" long			20	.400	Ea.
3000	Mirror with stainless steel, 3/4" square frame, 18" x 24"			20	.400	Ea.
3100	36" x 24"			15	.533	Ea.
3200	48" x 24"			10	.800	Ea.
3300	72" x 24"			6	1.330	Ea.
3500	Mirror with 5" stainless steel shelf, 3/4" sq. frame, 18" x 24"			20	.400	Ea.
3600	36" x 24"			15	.533	Ea.
3700	48" x 24"			10	.800	Ea.
3800	72" x 24"			6	1.330	Ea.
4100	Mop holder strip, stainless steel, 6 holders, 60" long			20	.400	Ea.
4200	Napkin/tampon dispenser, surface mounted			15	.533	Ea.
4300	Robe hook, single, regular			36	.222	Ea.
4400	Heavy duty, concealed mounting			36	.222	Ea.
4600	Soap dispenser, chrome, surface mounted, liquid			20	.400	Ea.
4700	Powder			20	.400	Ea.
5000	Recessed stainless steel, liquid			10	.800	Ea.
5100	Powder			10	.800	Ea.
5300	Soap tank, stainless steel, 1 gallon			10	.800	Ea.
5400	5 gallon			5	1.600	Ea.
5600	Shelf, stainless steel, 5" wide, 18 ga., 24" long			24	.333	Ea.
5700	72" long			16	.500	Ea.
5800	8" wide shelf, 18 ga., 24" long			22	.364	Ea.
5900	72" long			14	.571	Ea.
6000	Toilet seat cover dispenser, stainless steel, recessed			20	.400	Ea.
6050	Surface mounted			15	.533	Ea.
6100	Toilet tissue dispenser, surface mounted, S.S., single roll			30	.267	Ea.
6200	Double roll			24	.333	Ea.
6400	Towel bar, stainless steel, 18" long			23	.348	Ea.
6500	30" long			21	.381	Ea.
6700	Towel dispenser, stainless steel, surface mounted			16	.500	Ea.
6800	Flush mounted, recessed			10	.800	Ea.
7000	Towel holder, hotel type, 2 guest size			20	.400	Ea.
7200	Towel shelf, stainless steel, 24" long, 8" wide			20	.400	Ea.
7400	Tumbler holder, tumbler only			30	.267	Ea.
7500	Soap, tumbler & toothbrush			30	.267	Ea.
7700	Wall urn ash receiver, recessed, 14" long			12	.667	Ea.
7800	Surface, 8" long			18	.444	Ea.
8000	Waste receptacles, stainless steel, with top, 13 gallon	↓	↓	10	.800	Ea.
8100	36 gallon			8	1.000	Ea.

Figure 6.13

Using the man-hour productivity method and the data from Figure 6.16, the activity time can be calculated as:

$$\frac{22\ \text{Bathrooms/Floor}}{8\ \text{Hrs./Day}} \times 4.49\ \text{M.H./per bathroom}$$

Not All Scheduled Work Time Is Production Time   Looking at the activity in the sample network for gypsum wallboard installation, it can be seen that hanging, taping, and drying are included. The man-hour productivity is .017 man-hours per S.F. (see Figure 6.14).
The time required to install the gypsum wallboard is:

$$\frac{395\ \text{M.H.}}{3\ \text{crews} \times 2\ \text{people/crew} \times 8\ \text{hrs./day}} = 8.2\ \text{days}$$

This allowance does not, however, take into account the time required to dry the mud used in the taping process. This drying time will not show up in any references on man-hour productivity as it does not involve actual work, but it must be added by the scheduler when the overall activity time is established. In this case, it would be reasonable to add three days for drying time, resulting in a total activity time of 12 days. This figure would be used as the scheduled time.

Man-Hour Productivity Does Not Govern Activity Time   Some types of work are not purely a function of the number of man-hours required per unit of work. This fact must be taken into account when scheduling additional crews or workers. This point is illustrated in the previous example of bathroom accessories installation. Increasing the number of carpenters decreases the activity time proportionally. One carpenter requires 12.4 days, two carpenters need 6.2 days, three carpenters take 4.1 days, etc. In this case, one carpenter is essentially independent of the others and work proceeds according to the basic productivity rate. However, not all types of work fit this situation, as in the case of spraying the acoustical ceiling. A *Lath and Plaster* page from *Means Man-Hour Standards*, 2nd edition (Figure 6.12) shows that the basic crew for spraying the acoustical ceiling is not a single person. Experience tells us that the rate of production in this situation is dependent on the machine rather than the number of workers. Adding one per person to the crew is, therefore, not likely to speed up production. Changes in scheduled times would have to be planned on the basis of adding subsequent crews of three people each, not just individual workers, as was the case with bathroom accessories. In order to vary the scheduled time, a manager would need to go up to six workers and two machines, or up to nine workers and three machines.

Applicability of Data Sources   Any calculation of activity time is only as good as the data that goes into the process. Fortunately, calculations of activity times need not be as precise as estimates, but the same points are valid nonetheless. Specifically, the scheduler must always ask the question: does this data apply to the situation at hand, and if not, how must it be altered by judgment in order make the calculation valid?

Learning Curves   Frequently, work is scheduled as a series of similar activities. For example, the footing work in the sample hotel could easily be scheduled as a series of groups of footings, say five at a time. This being the case, crews assigned to the footings would probably produce the last group much more efficiently than the first. This rising productivity or production rate i.e., the learning curve, should be reflected in the times provided by

092 600		Gypsum Board Systems	CREW	MAKEUP	DAILY OUTPUT	MAN-HOURS	UNIT
608	2051	Drywall, 5/8" thick, on walls, standard, taped and finished	2 Carp	2 Carpenters	940	.017	S.F.
	2100	Fire resistant, no finish included			1,700	.009	S.F.
	2150	Taped and finished			940	.017	S.F.
	2200	Water resistant, no finish included			1,700	.009	S.F.
	2250	Taped and finished			940	.017	S.F.
	2300	Prefinished, vinyl, clipped to studs			1,050	.015	S.F.
	3000	On ceilings, standard, no finish included			1,175	.014	S.F.
	3050	Taped and finished			730	.022	S.F.
	3100	Fire resistant, no finish included			1,175	.014	S.F.
	3150	Taped and finished			730	.022	S.F.
	3200	Water resistant, no finish included			1,175	.014	S.F.
	3250	Taped and finished			730	.022	S.F.
	3500	On beams, columns, or soffits, standard, no finish included			650	.025	S.F.
	3550	Taped and finished			450	.036	S.F.
	3600	Fire resistant, no finish included			650	.025	S.F.
	3650	Taped and finished			450	.036	S.F.
	3700	Water resistant, no finish included			650	.025	S.F.
	3750	Taped and finished			450	.036	S.F.
	4000	Fireproofing, beams or columns, 2 layers, 1/2" thick, incl finish			330	.048	S.F.
	4050	5/8" thick			300	.053	S.F.
	4100	3 layers, 1/2" thick			225	.071	S.F.
	4150	5/8" thick			210	.076	S.F.
	4600	Blueboard, 1/2" thick, standard, not incl. skim coat			1,800	.009	S.F.
	4650	Fireproof			1,800	.009	S.F.
	4700	5/8" thick, fireproof			1,700	.009	S.F.
	5050	For 1" thick coreboard on columns			480	.033	S.F.
	5200	For high ceilings, over 8' high, add			3,060	.005	S.F.
	5270	For textured spray, add			1,450	.011	S.F.
	5300	For over 3 stories high, add per story			6,100	.003	S.F.
	5400	For skim coat plaster, add	J-1	3 Plasterers 2 Plasterer Helpers 1 Mixing Machine, 6 C.F.	3,000	.013	S.F.
	5500	For acoustical sealant, add per bead	1 Carp	1 Carpenter	500	.016	L.F.
	5600	Sound deadening board, 1/4" gypsum	2 Carp	2 Carpenters	1,800	.009	S.F.
	5650	1/2" wood fiber	"	"	1,800	.009	S.F.
612		METAL STUDS, DRYWALL Partitions, 10' high, with runners					
	2000	Non-load bearing, galvanized, 25 ga. 1-5/8", 16" O.C.	1 Carp	1 Carpenter	420	.019	S.F.
	2100	24" O.C.			500	.016	S.F.
	2200	2-1/2" wide, 16" O.C.			410	.020	S.F.
	2250	24" O.C.			490	.016	S.F.
	2300	3-5/8" wide, 16" O.C.			400	.020	S.F.
	2350	24" O.C.			480	.017	S.F.
	2400	4" wide, 16" O.C.			390	.021	S.F.
	2450	24" O.C.			450	.018	S.F.
	2500	6" wide, 16" O.C.			360	.022	S.F.
	2550	24" O.C.			440	.018	S.F.
	2600	20 ga. studs, 1-5/8" wide, 16" O.C.			435	.018	S.F.
	2650	24" O.C.			510	.016	S.F.
	2700	2-1/2" wide, 16" O.C.			425	.019	S.F.
	2750	24" O.C.			500	.016	S.F.
	2800	3-5/8" wide, 16" O.C.			400	.020	S.F.
	2850	24" O.C.			480	.017	S.F.
	2900	4" wide, 16" O.C.			390	.021	S.F.
	2950	24" O.C.			450	.018	S.F.
	3000	6" wide, 16" O.C.			360	.022	S.F.
	3050	24" O.C.			440	.018	S.F.
	4000	LB studs, light ga. structural, galv., 18 ga., 2-1/2", 16" O.C.			425	.019	S.F.
	4100	24" O.C.			500	.016	S.F.

Figure 6.14

establishing slower times for early activities, and shorter durations for the later activities which are performed more quickly.

Activities with No Information Available Frequently a scheduler or project manager is faced with the task of putting together a schedule, and having little or no information on some of the activities. A prime example is the situation in which a general contractor is building a schedule which contains mechanical or electrical work, yet there is little or no expertise within the general contractor's firm in these fields. Under these conditions, the general contractor should take the following two actions.

First, it is essential that the subcontractors in question be consulted as soon and as often as possible. The subcontractors should be encouraged to supply activity time determinations based on the same methodology used by the general to determine times for other activities. Obtaining such information from subcontractors helps the scheduler set a more accurate activity time, and makes him aware of the manning levels needed by the subcontractor to meet his commitments.

Second, the general contractor must look at the role of each subcontractor in the overall picture. As the sample calculations are developed in the next section, it will be seen that while the mechanical and electrical times have not been calculated, times have been assigned. These times are based on the average activity times for the other, calculated activities, and on the subsequent, sound assumption that the mechanical and electrical subs must fit the overall sequence like everyone else. In this situation, it is common and effective practice to determine an overall timing for each phase, then go to the subs in question and say, this is the sequence and these are the time limits we all must meet; can you do it, and what will it take? The emphasis must be on consultation and cooperation with all parties if the schedule is to work effectively in a job environment.

Adjusting Activity Times

There are a variety of reasons for adjusting activity times. An example situation is shown in Figure 6.15. In this case, the sequencing for the various phases of interior work is performed as the building construction proceeds upward. As a result, overall activity times must be adjusted in conjunction with other dependent or related activities. It can be seen from this small sample network that the plan is to perform the first activity — erect door frames — on the first floor, then move the door crew up one floor while bringing in the second activity crew to install studs and furring in the area just vacated by the door frame crew, and so on up the building.

When it comes to calculating the days required for these activities, published cost/productivity data may have to be adjusted. For example, using typical crew sizes from *Means Building Construction Cost Data* to calculate the number of days required results in a wide variation of activity times — from 1.1 days to 40.3 days per activity. (See Figure 6.16.)

Clearly, a project manager setting up work crews for the hotel would not blindly follow the calculated times, but rather would try to even them out to reasonably similar rates of progress in order to attain a manageable production sequence. This is done by varying the crew size in order to "invest" the necessary number of man-hours per day to do the work in the allotted time. As an example, say the project manager had decided for reasons of overall project time that each floor needed to be done in 75 days working time. It can be seen that the overall process for each floor consists of 14 basic steps, and dividing this number into 75 days yields an average activity time of approximately five to six days. Using the activity of laying carpet (see Figure 6.16), which required 322 man-hours to accomplish and will take 40.3 days using the "average crew" of one carpet layer, the activity time can be adjusted to 6.7 days (rounded to seven days) by simply applying six carpet layers to the task instead of only one. This concept can also be illustrated using the following formulas.

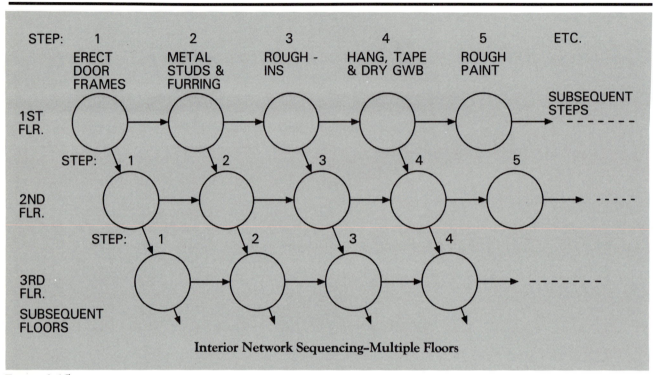

Interior Network Sequencing–Multiple Floors

Figure 6.15

Interior Network Time Calculations						
	Quan.		MH/ Unit	Total MH	Typ. Crew Size	Days Reqd.
Erect hollow metal door frames	46	ea	1.067	49	2	3.1
Install metal studs and furring	13,960	sf	.017	237	1	29.7
Hang, tape, & dry gypsum wallboard	23,232	sf	.017	395	2	24.7
Install ceramic tile	2,244	sf	.087	195	2	12.2
Spray acoustical ceiling plaster	13,433	sf	.046	618	3	25.7
Rough paint	16,940	sf	.011	186	1	23.3
Install bathroom counters and tops	154	lf	.800	123	2	7.7
Hang vinyl wall covering	4,048	sf	.013	53	1	6.6
Install ceiling grid	848	sf	.010	8	1	1.1
Layin acoustical ceiling tiles	848	sf	.012	10	1	1.3
Install bath and closet accessories	22	ea	4.487	99	1	12.4
Lay carpet	1,492	sy	.216	322	1	40.3
Hang doors and hardware	46	ea	1.000	46	1	5.8
Touch up painting	1	ls	16.000	16	2	2.0

Figure 6.16

$$\text{No. of man-hours per day(s)} = \frac{\text{Total man-hours required}}{\text{No. of days allowed}}$$

$$\text{Size of crew needed} = \frac{\text{No. of man-hours per day}}{8}$$

In the case of carpet laying, this formula yields:

$$\text{No. of man-hours per day} = \frac{322}{7} = 46 \text{ man-hours}$$

$$\text{Size of crew needed} = \frac{46}{8} = 5.8 \text{ carpet layers}$$

which would be rounded to 6 carpet layers.

This "averaging," or "rounding," of activity results in a much smoother and shorter overall project time, and is much easier for the project manager to control. It does, however, require considerable coordination with the subcontractors. They must not only understand their position in the project sequence, but also the necessity that they complete their work in the allotted time so that subsequent crews will not be delayed.

Calculating Overall Job Duration

After the times have been calculated for the individual activities, the scheduler or project manager is then in a position to determine how long the entire project should take to accomplish. This is done by applying the times or durations for the activities to the logic diagram, which determines the order in which the activities are performed. To illustrate this procedure, the sample foundation network is used. Included are activities from the beginning of mass excavation to clean-up and complete foundation. See Figure 6.17.

It should also be noted that the scheduling procedures are fairly straightforward and can be performed manually. The calculations can, however, be tedious, and it is easy to make detail errors. It is therefore recommended that anyone involved in scheduling construction projects understand the calculation procedures and perform them using a good CPM computer scheduling program.

Goals of the Project Calculation Procedure

The last of the basic steps in the planning and scheduling process is determining several facts about the upcoming work, based on the assumptions and plans built into the logic diagram and activity times. These goals are:

1. To determine the desired or possible starting times for each of the activities which have been identified and established as necessary for getting the whole job done.
2. To determine the finishing time for these activities.
3. To determine how much flexibility is possible in these start and finish times, given the constraints which have been identified as affecting the project.
4. To determine the activities which are crucial to the success of on-time performance, i.e., the critical path.

Once the above information is known, the project management team is in a position to effectively control the work of all the parties whose tasks must be performed in order to get the entire job done.

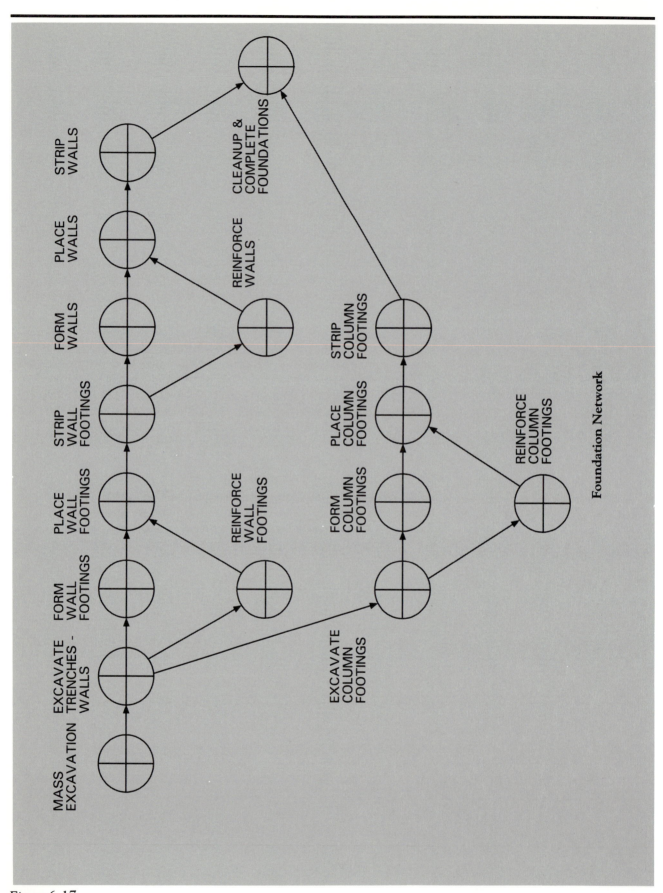

Figure 6.17

Foundation Network

90

Definitions

To determine the above information, it is necessary to first define some terms which are commonly used in the calculation process.

Succeeding Activity/Preceding Activity — Activity A precedes Activity B. Activity B is the succeeding activity. In logic terms, a preceding activity must be complete before a succeeding activity can begin.

Duration (DUR) — the calculated or estimated time for an individual activity (determined in the previous section of this chapter).

Early Start (ES) — the earliest possible time an activity can start, assuming all previous activities have been completed. The early start of an activity is also the early finish of the preceding activity.

Early Finish (EF) — the earliest possible time an activity can be completed, assuming again that all previous work has been completed. In practice, the EF is equal to the activity time added to the early start time.

Late Finish (LF) — the latest time an activity can finish, and still not delay the completion of the project as a whole. The late finish of an activity is equal to the late start of the succeeding activity.

Late Start (LS) — the latest time an activity can begin, and still not delay the final completion. In practice, LS is equal to the activity time subtracted from the late finish.

Total Float (TF) — the difference between the early and late times on an activity; it is the allowable delay in starting the activity.

Critical Path (CP) — the set or path of activities the early times of which are equal to their late times, i.e., total float equals 0. In other words, Critical Path represents the group of activities that must start on time in order to keep the project as a whole on time.

Milestone Activity — an activity with zero duration, which marks the end of a particular phase of work.

The Actual Calculation Procedure

The actual process used to determine early times, late times, and floats consists of three individual steps, which must proceed in order. None of the steps can be bypassed in order to obtain the information on the next one, as each depends upon its predecessor. These steps can be tedious, but are unavoidable. Specifically, they are:

1. *Forward Pass* — the procedure whereby the early times for a project are calculated. It is called the *forward pass* because it proceeds "forward" along the logic diagram from left to right.
2. *Backward Pass* — the procedure in which the late times of a project are determined. It is called the *backward pass* because it proceeds "backward" along the logic diagram from right to left.

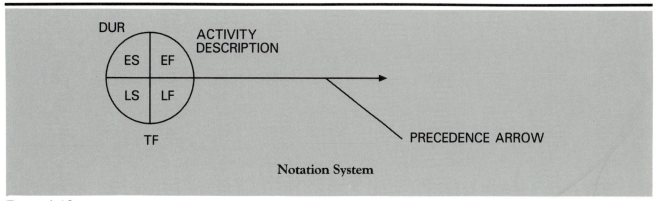

Figure 6.18

3. *Calculation of Floats and Critical Path* — the procedure in which the difference between the allowable starting times is calculated and the activities which have no slack are identified.

Demonstration of the Actual Procedure

The entire project calculation procedure is demonstrated step-by-step in the following figures. The various times for each activity are shown using the notation system described in Figure 6.18.

Forward Pass The forward pass is illustrated in Figure 6.19. It begins on the left side of the sample logic diagram, in this case with the initial activity of mass excavation. Normal practice is to begin any network with a single starting activity. This is not absolutely necessary, but it simplifies the calculation procedure considerably.

To start the process, mass excavation is assigned an early start time of Day 0. The early finish of this activity can then be calculated as E6 + DUR = EF, or 0 + 7 = 7. These numbers are then written in the appropriate quadrants of the node, according to the notation system shown in Figure 6.18.

If we define *early start time* as the earliest time any activity can begin assuming all previous work has been completed, it can be deduced that the early start time for the next activity, *excavate trenches*, is Day 7. In other words, excavating for the wall footings trenches cannot be started until mass excavation is complete; the major excavation is complete on Day 7; trenches start immediately thereafter.

Using the same logic as that applied to mass excavation, it can be seen that *excavating trenches — wall footings* can finish on Day 8 (7 + 1 = 8). Further, looking at the next phase of the sample network, there are three activities that cannot start until wall trenches are complete. These are *form wall footings, reinforce wall footings,* and *excavate trenches — column footings.* All of the activities at this stage can start immediately after *excavating trenches — wall footings.* Therefore, their early starts are Day 8.

Each one of these activities has a different early finish, however, since each has a different duration. For example, *forming wall footings* can be completed at Day 12, but *reinforcing wall footings* can be completed much earlier due to its shorter duration of one day. Similarly, *excavating for column footings* can also be completed on Day 9.

Looking ahead to *placing wall footings,* it can be seen that this activity is dependent on (i.e., preceded by) two activities. Each of these activities has a different early finish time, so the question arises — what is the earliest possible time *placing wall footings* can begin? The answer lies in the definition of early start — the earliest time an activity can begin assuming *all* previous work has been completed. Under this definition, *placing wall footings* cannot begin until *forming wall footings* has been completed, i.e., Day 12. Concrete placement cannot begin at Day 9 with the completion of reinforcing, because at that time, *forming wall footings* is still underway, with three more days to go, and the logic diagram shows a dependency between forming and placing.

A simple rule of thumb for dealing with multiple preceding activities is to take the early start as the last early finish of all preceding activities. Using this rule, the choices are 12 and 9. The early start of *placing wall footings* becomes 12, since it is the last of the preceding activities to finish.

Continuing across the short diagram, it can be seen that the last activity, *clean-up and complete foundation,* which is a milestone activity with zero duration, can occur on Day 39. This means that for this sort sample diagram, the earliest possible finish for the "project" is Day 39.

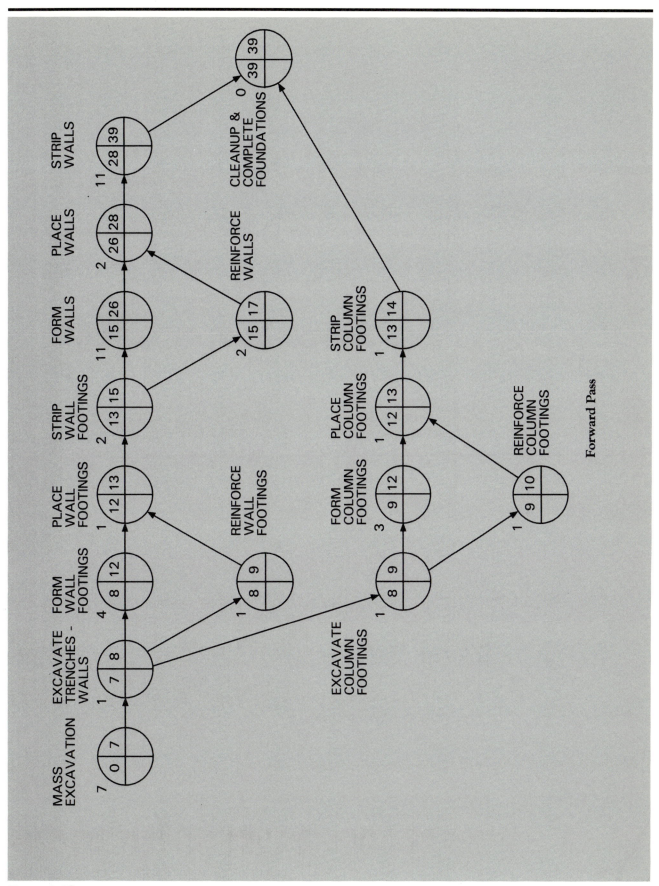

Figure 6.19

At this point, the forward pass is complete, and all activities have had their early start and finish times calculated. While the calculation process is by no means complete, the diagram does have value as a schedule at this point. For example, if the mass excavation had just begun, it would be possible to tell the concrete placement contractor this his crews would probably need to be on the job approximately 12 days hence.

Backward Pass

The backward pass is illustrated in Figure 6.20. It begins on the right side of the logic diagram, in this case with the last activity of *clean-up and completion* of the foundation. Following the definition of late finish, this last activity can be assigned a late finish of Day 39. Logically, this judgment is based on the fact that if it finishes any later than Day 39, the project as a whole finishes late. A simple rule of thumb is that it is common practice to have a single ending node for a network and the late finish is equal to the early finish for that node.

To determine the late start for *clean-up and complete foundation,* the definitions are called on once again. In this case, the late start of the last activity is equal to the late finish minus the activity duration $(39 - 0 = 39)$. This is a logical progression since if clean-up and completion must finish by Day 39, and these milestone activities require zero days, then they must start by Day 39 in order to finish on time.

Looking back across the diagram, the activities which precede *clean-up and completion* are *strip walls* and *strip column footings.* The question arises as to how late these activities can finish without delaying the start of the clean-up. The answer is that they must finish by the time of the latest start of *clean-up and completion,* i.e., Day 39. The rule for establishing late finishes of preceding activities is that they are equal to the late start of the following activity.

Again, proceeding leftward and calculating the late start of *strip walls* is a matter of subtracting the activity time of eleven days from the late finish, establishing the late start as Day 28: $(39 - 11 = 28)$.

Looking at the diagram, it can be seen that *placing walls* is preceded by two activities. It can also be seen that both must be completed by Day 24 in order for *placing walls* to begin by the late start time of Day 24.

Looking further left, the late starts of each of these activities can be calculated by subtracting the activity duration from the late finish of Day 26. This calculation produces the following late starts on these two activities: 15 and 24.

The next question involves determining the latest possible finish for *stripping wall footings* such that it will not delay the start of the wall forming and reinforcing activities and foundation as a whole. The answer can be found again in the logic of previous definitions. The late finish is the latest time at which *stripping wall footings* can begin without delaying the following activity. To illustrate the point, take the late start for *reinforcing walls* and ask the question, if *wall stripping* is completed at Day 20, what happens to the start of subsequent activities? To answer the question, consider the effect on *forming walls.* If *stripping wall footings* is not finished until Day 20, then *forming wall footings* could not start until Day 20, which is five days past its already calculated late start time of Day 15.

The simplest rule of thumb in dealing with multiple succeeding activities is to take the lowest, or earliest late start among the succeeding activities as the late finish for the preceding activity. In this case, it is the late start of 15, which is the lowest of the two forming and reinforcing activities, which becomes the late finish of *stripping wall footings.*

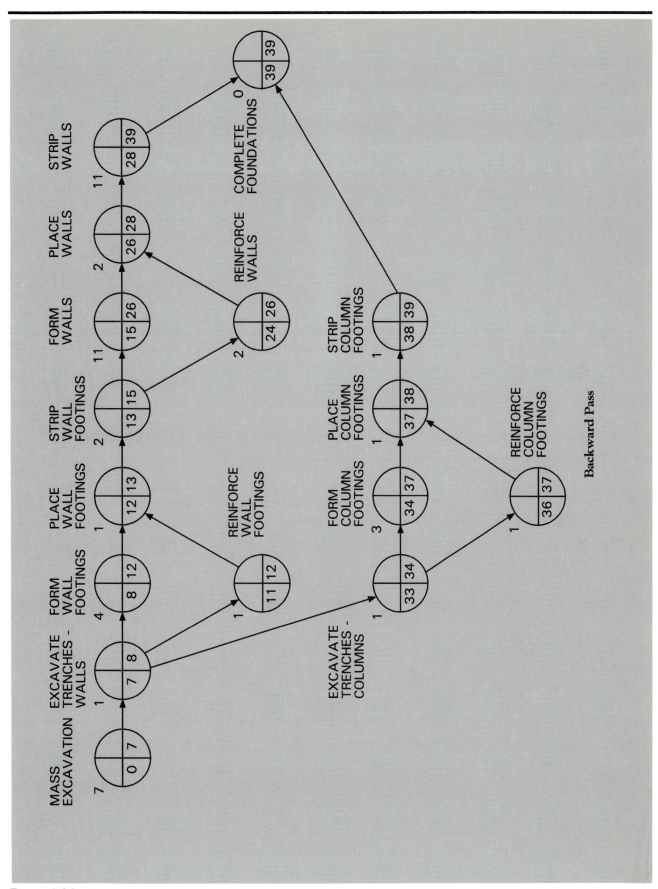

Figure 6.20

Backward Pass

95

Continuing to the left of the diagram, the backward pass is completed at the beginning node with a late finish (for *mass excavation*) of Day 0. This determination is consistent with the logic that the late start is the latest time that the activity can begin without delaying any other activity. In fact, in networks with single beginning and ending nodes, the calculation of the backward pass must result in a late finish of 0; otherwise, there may be an error in the calculation.

Calculation of Floats and Critical Path   Now that the forward and backward passes have been completed, and the early and late times determined, the total float for all activities can be calculated. This part of the procedure is much simpler than the last two steps, and consists only of subtracting the early start time from the late start time (or early finish from late finish) for each activity, in turn.

Looking at Figure 6.21, it can be seen that the TF (Total Float) for *mass excavation* is $0 - 0 = 0$; for *excavate trenches*, it is $7 - 7 = 0$. Looking at the framing and reinforcing activities, however, it can be seen that while the float for *forming wall footings* is also 0, other activities have different values. *Reinforcing wall footings*, for example, as a TF of $11 - 8 = 3$ days. A value for float of more than 0 means that these activities have some "slack," or permissible delay, built into the times that they can start. In other words, *reinforcing footings* has a "window" of Day 8 to Day 11; it can start anytime during this period and still meet the requirements of the project as a whole.

Further, it can be seen that the activities which have 0 float, (i.e., *mass excavation, excavating trenches — wall footings, forming wall footings,* etc.) constitute the critical path for the project. These activities must start at their scheduled early times. Otherwise, the project will be delayed beyond the calculated completion time of Day 39.

Advanced Calculations

As noted earlier, the basic relationship between activities is the *finish to start*, or *FS*, relationship. This relationship is by far the most common type used in networks, and was used in the previous section covering basic forward pass and backward pass calculations. The other relationships, *start to start* (SS), and *finish to finish* (FF), are also very valuable. Their calculation is as follows.

Start to Start   In calculating the SS relationship, rather than basing the start time of the succeeding activity on the finish of the preceding one, each start time is based on the start time of the preceding activity. This concept is best shown in Figure 6.22, which also shows the notation system and time scale of this relationship. In this diagram, two activities, A & B, have times of 10 days and 20 days respectively, and are connected by an SS 5 relationship. In order to calculate the start time of the succeeding activity B, the lag value of 5 is added to the start time of 0 for the preceding activity A, to arrive at a start time of 5 for the succeeding activity B.

Finish to Finish   The calculation of the FF (Finish to Finish) relationship is similar in concept, though it is slightly more complex. To illustrate this point, Figure 6.23 shows two activities A & B, which have times of 20 and 15 days respectively, and are connected by an FF 5 relationship. In order to calculate the starting time for B, the early finish for A must first be determined; it is $0 + 20 = 20$. The finish time for B can then be calculated by adding the lag value of 5 to the finish time of A, or $20 + 5 = 25$. Now that the finish time of B has been determined to be 25, the early start of B can be determined by subtracting the duration of Activity B from the finish time of B, or $25 - 15 = 10$.

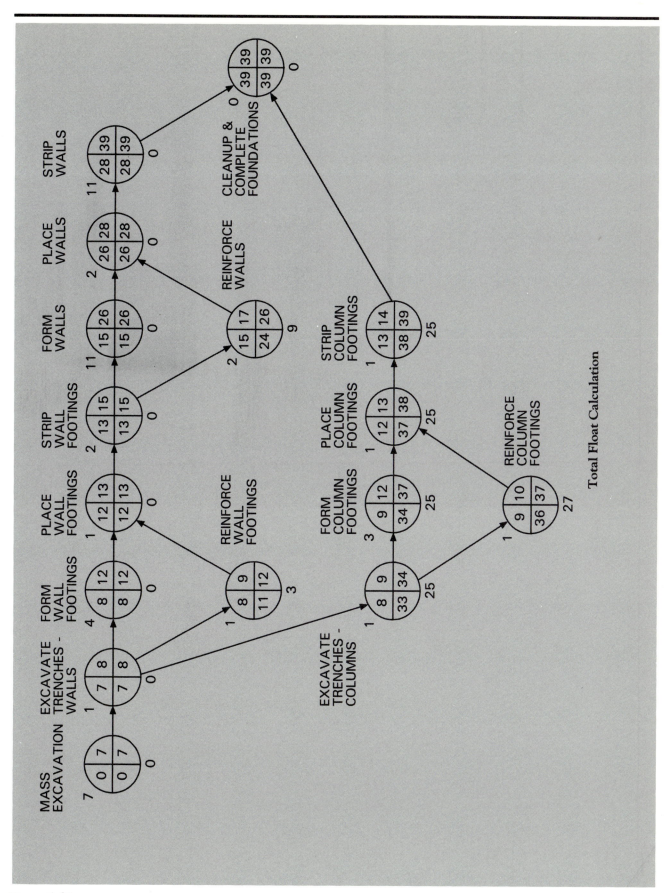

Figure 6.21

Total Float Calculation

97

Constrained Dates

In addition to showing overlapping relationships, the scheduler is often faced with the problem of showing or factoring in events or circumstances which are outside the actual construction process.

For example, it is very common in construction to have delivery dates determined by factors completely outside the job site, such as manufacturing

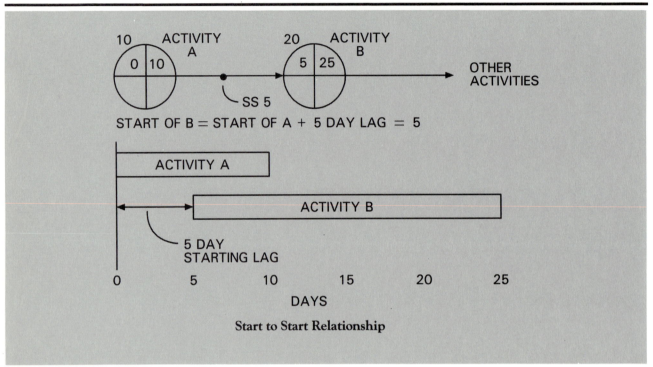

Figure 6.22

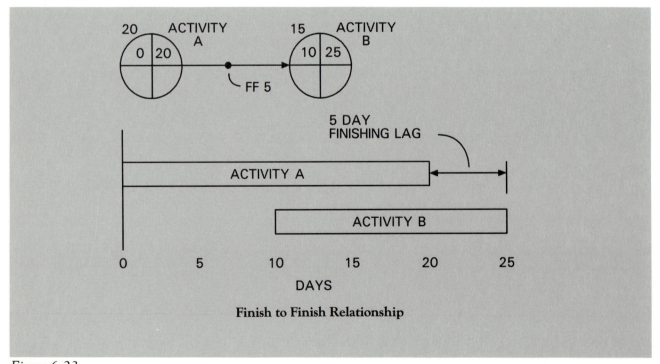

Figure 6.23

schedules. In another typical circumstance, a portion of the site may not be available until a given date, or an owner may impose a required finish date on all or part of the project. In these cases, it is common practice to use what are known as *constrained dates,* or as they are sometimes known, *plugged* dates. These constraints are classified as *no later than* (NLT) or *No Earlier Than* (NET) dates, and each affects the schedule calculations differently.

The NET date affects only the forward pass. To illustrate the point, Figure 6.24 shows the small sample network with a constraint added. Specifically, the constraint is that the gypsum wallboard cannot be delivered to the job site until Day 25. Logic tells us that the gypsum wallboard installation cannot take place before the material arrives—a fact that must be considered in the calculations. This situation leads to what is known as a *Start No Earlier Than* (SNET) of Day 25 for gypsum wallboard. The SNET is determined by simply imposing the constraint of Day 25 over the calculated day of 20.

Continuing the calculations for the rest of the forward pass shows an early finish for gypsum wallboard of 34, and ES/EF (Early Start/Early Finish) of 34 and 40 for rough paint. This schedule puts the completion of the project at Day 40, which is five days later than originally planned, and equal to the five day delay in the start of the gypsum wallboard caused by the delay in delivery.

It must be noted however, that if the constraint was a SNET Day 18, it

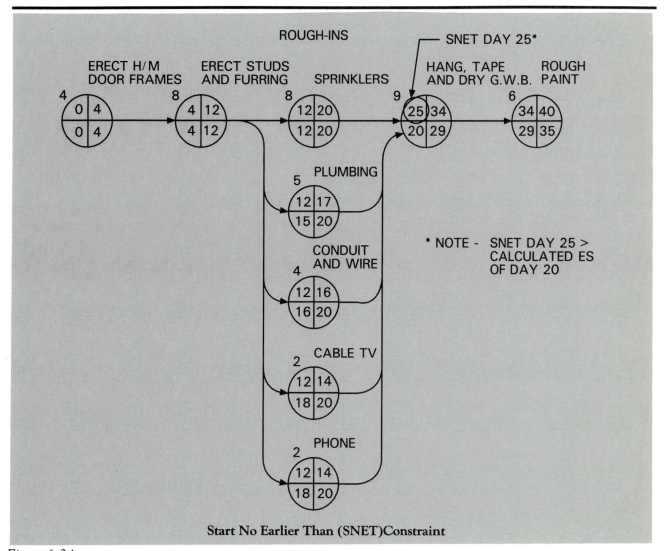

Start No Earlier Than (SNET)Constraint

Figure 6.24

would have no effect, since the gypsum wallboard could not start until Day 20 anyway.

By contrast, the NLT (Not Later Than) date affects only the backward pass. As an example, Figure 6.25 points out a case in which, for some external reason, the rough paint must be finished no later than Day 32 (FNLT Day 32). In this case, the constraint is simply imposed over the late finish of rough paint, which becomes Day 32 instead of Day 35. The backward pass calculation is then carried out as before.

Note that at this point, the late start time for hollow metal door frames is now −3. This is not an error; the reason for this figure can be shown if the float calculations are carried out. The floats now show a critical path with floats all equal to −3. This float is now what is known as *Negative Float*; it tells the project manager that this project is three days behind before it even starts, if the constraint of finishing no later than day 32 is to be met. It means that somewhere along the critical path, some times must be reduced in order to meet the schedule requirement. Finally, as with the SNET constraint, it should be noted that the constraint would not affect any other activities.

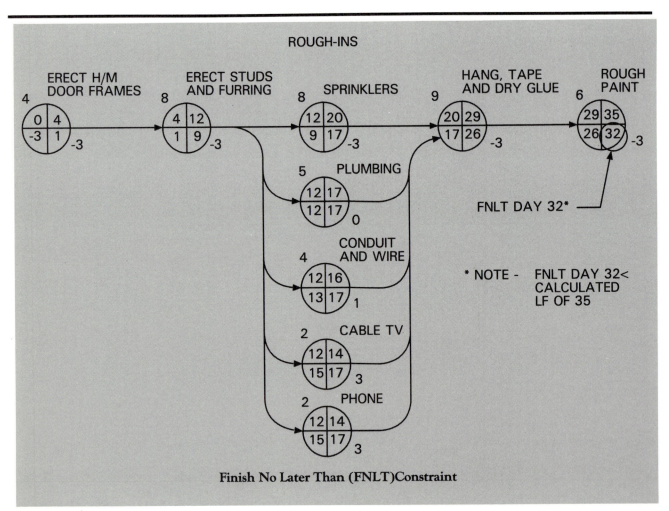

Finish No Later Than (FNLT)Constraint

Figure 6.25

100

It is possible to have *finish no earlier than* constraints, and *start no later than* constraints, but these situations do not commonly occur. In any case, the effect on calculations is the same.

Free Float

In addition to total float, there is *free float,* which is defined as the differences between the early finish time of a preceding activity and the early start time of the succeeding activity.

Free float defines the amount of delay that can be allowed without affecting any other activities in the project. Free float is always more restrictive than total float. Frequently, an activity may have some total float and no free float. Free float is more difficult to calculate by hand, and is generally not done in manually calculated networks.

Conversion of Work Days to Calendar Dates

As noted earlier in this chapter, the calculations preformed were in work days. There are very good reasons for using work days as a time unit when calculating schedule information. However, use of the work day unit raises some problems in the *use* of schedules. Information on the scheduling process must be communicated to the persons on the job site who are responsible for carrying out the work. Unfortunately for the person who calculated the schedule in work days, no one in the real world thinks in these terms. For example, to tell the mechanical subcontractor that Rough-in HVAC starts on Day 12 really tells him nothing. The calculated work day schedule must be converted to calendar days, meaningful to the users of this information.

Basic Calendars   The easiest conversion technique is to simply assume a five-day work week, and then divide the number of work days by five; then multiply by seven days per calendar week, and count the elapsed calendar days.

For example, if the Rough-in HVAC activity of the sample project were to start on August 31, 1987, which is a Monday, then conversion could be made by dividing 12 by 5 work days per week, then multiplying by 7, which would equal 16.80 days, rounded to 17. By counting the elapsed calendar days from the start date of August 31, the starting date of rough-in HVAC is established as September 16. This method is, however, quite cumbersome and inexact (note the calculated figure of 16.8 calendar days) and does not take into account holidays and other non-work periods.

A much better method is to establish a project calendar which defines the working days of the job, and assigns a date to each working day. This is done by drawing up a calendar similar to the one shown in Figure 6.26. The technique involves first defining the start date of the project (August 31 in the example), and then counting work days forward, skipping holidays and non-work periods when appropriate. In the example of Figure 6.26, it can be seen that during the months of September and October, September 7 (Labor Day) and October 12 (Columbus Day) are holidays and are therefore bypassed in the work day sequence. The finish date of studs and furring can be identified on the calendar; the start date of rough-in HVAC can be established as the beginning of the next day, September 16.

Clearly, the above information is detailed and may involve tedious computations. This is the sort of data which can easily be computerized and would benefit from this approach. The fact is that all available CPM computer systems are designed to allow the establishment of at least one basic calendar — one that takes into account all holidays, and will automatically convert and display as calendar dates the results of the activity data input to

DATE → 28
WORKDAY → 20

SEPT 87 / OCT 87

S	M	T	W	TH	F	S
30	31 — 1	1 — 2	2 — 3	3 — 4	4 — 5	5
6	7 — H	8 — 6	9 — 7	10 — 8	11 — 9	12
13	14 — 10	15 — 11	16 — 12	17 — 13	18 — 14	19
20	21 — 15	22 — 16	23 — 17	24 — 18	25 — 19	26
27	28 — 20	29 — 21	30 — 22	1 — 23	2 — 24	3
4	5 — 25	6 — 26	7 — 27	8 — 28	9 — 29	10
11	12 — H	13 — 30	14 — 31	15 — 32	16 — 33	17
18	19 — 34	20 — 35	21 — 36	22 — 37	23 — 38	24
25	26 — 39	27 — 40	28 — 41	29 — 42	30 — 43	31

Workday to Calendar Date Conversion (5 Day Week)

Figure 6.26

the system. Many systems allow the creation of multiple calendars, some with varying holidays and varying work weeks, and permitting the assignment of activities to a different calendar to suit the requirements of the job. The availability of these features provides yet another strong argument for the purchase and application of good computer software for scheduling.

Dealing with the Effect of Weather on Project Calendars There are several ways to deal with the weather factor when scheduling construction jobs. The first principle is that only those activities that are affected by weather should be adjusted or adapted. For example, weather clearly has an effect on excavation, but has no impact on the installation of bathroom tile. Also, some activities can be performed as quickly in poor weather as in good if the proper precautions are taken. Concreting is an example for which weather becomes a cost concern, but not necessarily a schedule concern. The key is to treat each activity individually according to its particular requirements.

There are two methods for dealing with the effect of weather on activity duration; these are: 1) add time to the activity to compensate, or 2) schedule fewer working days within the calendar period. To illustrate the point, consider an excavation activity which is scheduled to take 20 days. If records show that it rains an average of 25% of the days in the season and location specified, then the activity length can be increased to 25 days. The alternative is to schedule the activity on the basis of 4-day rather than 5-day weeks. In either case, the activity is effectively "stretched" by 25%, thus providing extra time to make up for those days when no excavation work can be performed. Lengthening the activity time is the simpler of the two methods, though it will not work in certain circumstances. In such cases, multiple calendars may be needed to reflect the differing effects of weather on various activities. This multiple calendar approach would be needed when it is determined that no work of a certain type could be performed during a certain season, or seasons. A specific example might be the erection of structural steel in Wyoming which may be impossible during the months of December, January, and February. A calendar would have to be established to accommodate the steel erection during the appropriate season(s).

7

Management of Submittal Data and Procurement

Management of Submittal Data and Procurement

The emphasis so far has been on planning and scheduling the physical building tasks necessary to complete a project. The construction process involves more, however, than the actual placement of concrete, steel, and other materials. The technical complexity of many of today's construction projects requires the purchase of materials meeting rigorous specifications, often from remote vendors. Contractors must frequently go through a long review and approval process for these materials, submitting technical data to the owner and/or architect/engineer for determination of a product's suitability for the project. Only after this review and approval can the actual purchase and installation take place.

An entire set of administrative procedures has grown up around this purchasing requirement, which is typically called *submittal data/shop drawing and procurement.*

Submittal data and procurement are proving to be major problems for many contractors. It is an unavoidable process, but one that is often a significant cause of delay in the overall building process. This chapter provides guidelines for the project manager who is interested in setting up a simple, yet effective system. Using these prescribed methods, submittal data and procurements may be tracked and reported accurately. With access to reliable data, potential problems can be recognized and addressed early, thereby minimizing delays.

The Source of the Problem

Clearly, construction work cannot be performed unless the necessary materials are delivered to the job site on time. For readily available materials, this is usually not much of a problem. In the case of more unusual materials, however, and most equipment, the review, approval, and procurement process may create the need for a long lead time. Why are so many delays caused by procurement problems? The reasons are many and varied; some of the most prominent are listed below.

First, the procurement process is inherently cumbersome and involves a considerable amount of detailed record keeping and paperwork. Consequently, there is ample opportunity for error and oversight.

Second, most construction professionals are oriented not to the administration of paperwork, but rather to physical action. Many of their careers have been built around the management of actual construction tasks, not "office work." It is only natural for such individuals to concentrate on the problems in the field, paying less attention to off-site planning.

Third, most of the tasks necessary for good procurement management take place off the site. As a result, the project management team may tend to take an "out of sight, out of mind" attitude. Unless a distinct effort is made to do the necessary administrative tasks, the progress of the procurement process is not visible and is therefore ignored.

Fourth, many parts of the process are controlled by others who may not share the project manager's sense of time, schedule, and priorities. For example, the architect/engineer is not contractually obligated to the contractor in any way. He is acting as the agent of the owner, and as such has a limited legal obligation to act expeditiously in carrying out the review tasks required of him. The subcontractors, while legally obligated to the contractor, often do not have the same priorities as the general contractor, and may have equally pressing obligations to other projects. The subcontracting firms may be manned by personnel even less office-oriented than the general contractor, further compounding the "out of sight, out of mind" point of view. Finally, suppliers of materials and equipment have their own schedules for production — schedules that may have only the remotest connection with the general contractor's project. Their production schedules are typically determined by the order in which the materials required for their products are purchased, leaving the contractor with little, if any, control over the fabrication schedule which affects his specific project.

Procurement is an essential process. It has the potential for causing significant delays, yet the project manager has only limited control over many parts of the process. It is therefore critical that the project manager pay extra attention and take every action to overcome the problems, thereby ensuring a smooth, steady flow of materials to the job.

Basic Procurement Procedures

The basic procurement procedures are typically specified by the architect/engineer and detailed in the project specifications. These procedures typically resemble those outlined schematically in Figure 7.1.

Generally, the project manager begins by issuing a purchase order or subcontract to the appropriate supplier or subcontractor. The supplier or subcontractor then prepares what is known as *submittal data*, which is transmitted to the contractor at the job site. This submittal data can take many forms. One common type of submittal is shop drawings, engineered drawings that further amplify the working drawing for the project. Also, suppliers typically provide *catalog cuts*, which are simply pages from their catalogs that provide technical information.

Certificates of compliance or inspection reports may also be required. In any case, the exact requirements should be stated in the project specifications. After a general review by the project manager or other member of the project team, the requirements are transmitted to the owner and architect/engineer for review and approval. Once approved, the submittal documents are transmitted back down the chain to the appropriate parties, who then carry out the tasks of fabrication and delivery. Only after all this has taken place can the material be installed on the project.

Key Elements in Successful Procurement

Procurement is a straightforward process in most cases. If the quantities of data transmitted are small, there should be few, if any, problems in carrying the procurement out effectively. However, the amount of data on a typical job is usually very large. To manage procurement activities effectively, the project manager must keep certain key principles in mind, and adapt the process to the specific job at hand.

First, it is vital to identify not only the firm (subcontractor or supplier) used at each stage of construction, but also a personal contact at each firm. For example, the architect and/or engineer's offices generally assign project captains to monitor the designer's reviews. Each subcontractor assigns a project manager; each supplier assigns a salesperson. The project manager (or whoever he assigns to the task of tracking submittal data) should establish a personal working relationship with these representatives and maintain regular contact with them, keeping them informed of concerns and problems at the job site. This contact will in many cases be daily.

Second, a good tracking system must be maintained. The purpose of this system is to provide the project manager with a record of each piece of submittal data — including where it is at any time in the process. In this way, the project manager can follow up on those items that are not progressing through the process quickly enough, and take action to ensure timely delivery of material to the job. This system also allows the project manager to find out which parties are consistently holding up the process. In the event of a claim for extension of time or other compensation, the project manager can use the tracking records to make a case concerning others causing delays to the job.

Third, communication and follow-up are crucial. For example, it does no good for the project manager to know that the architect is holding up several key shop drawings if no one calls the architect's project captain to promote action on the reviews. Submittal data/procurement probably involves more parties than any other aspect of the overall construction process, and communication problems always increase as the number of communicants

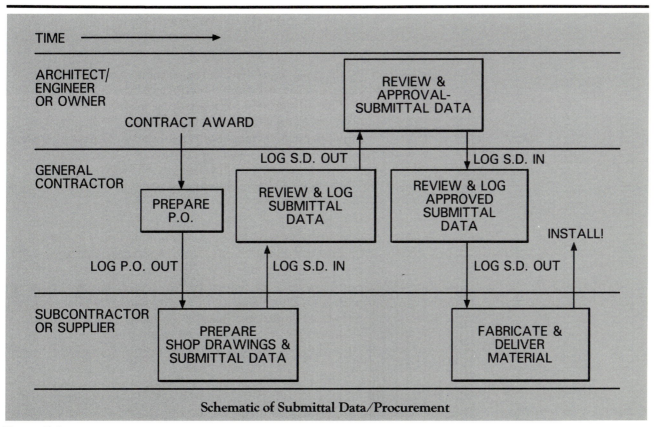

Schematic of Submittal Data/Procurement

Figure 7.1

109

grows. In addition to personal contacts with the various parties, it is also an excellent idea to include detailed reviews of submittal data status in the weekly or monthly job schedule meetings. At these meetings, responsibility can be assigned to project team members for tracking down delayed submittals and deliveries (as identified in the meetings).

Record Keeping and Tracking

The first task in the management of the procurement process is setting up an efficient record keeping system. A good system can be maintained easily by the project team members and provides data about the mass of submittals to be developed and submitted during the course of construction. Setting up and maintaining this system is basically a three step procedure:

1. Making a list of the items that require submittal data;
2. Keeping records of their progress in a log;
3. Coordinating the submittal data processing with the construction schedule.

Making a List of Submittal Items

The first and probably the most difficult task is determining all of the documents that must be submitted to the owner and architect for approval prior to purchasing the material. Unfortunately, this list will probably be voluminous on all but the smallest jobs. It takes some time to put such a list together; the day after the contract award is not too soon to begin. The starting point for this task is the *project specifications*. If the architect and engineers have done their job properly, each required submittal will be listed in the beginning of each specification section or sub-section. Ideally, the specifications should list exactly what documentation is required. If this is not the case, some judgment based on experience is required of the person developing the actual list. Ideally, development of this list is best assigned to one person in order to prevent gaps. However, on some jobs, preparing a list of submittals may be too large a task for one person to finish in a reasonable period of time.

Once a comprehensive list of submittal items is drawn up, it must be determined exactly who is responsible for the preparation of each item. Most of the submittal items come to the general contractor from subcontractors. If the subcontractor has an organized and professional management system, the general contractor can probably expect submittal items to be delivered on time. If they are not on time, there is a need for follow-up. The person writing the submittal list must be aware of which subcontractor has the contract for the various elements of each section. For example, while Division 15 includes both plumbing and HVAC systems, these two elements are typically performed by separate contractors.

Once the list is developed by the general contractor, the various sub-lists should be transmitted to the persons responsible for submitting the data. It is a good idea to present these lists to each subcontractor in an initial meeting. At this time, the subcontractor may be asked for commitments on submittal times and delivery. Based on this information, the project manager can get a better feel for whether or not any of the subcontractor's material deliveries can be expected to cause delay problems. The contractor should also make clear to all subcontractors and suppliers the *procedures* for submitting data. (For example, should the material be sent to the home office or to the job site?)

Finally, it is impossible to be too thorough in drawing up a list of submittal data items. Nothing is more frustrating to a project manager than to have a job going along smoothly only to discover that everything must come to a halt because a key piece of equipment has not arrived.

Keeping a Log of Submittal Data Approvals

Once the list of items has been developed and all parties are aware of their responsibilities for submittal and procurement, the submittal items begin to come to the job site for review. The project manager must ensure that detailed and thorough records are kept of the progress of items up and down the chain (See Figure 7.1). The easiest and best method involves maintaining a log similar to that shown in Figure 7.2. This example is not the only acceptable format; other possibilities for workable log formats are shown in Figures 7.3 through 7.5.

The procedure for record keeping is quite straightforward: each time a document passes through the project manager's office, it is logged in the appropriate column. As shown in Figures 7.1 and 7.2, the log is simply a recording of actions taken by the project team members on the job. For example, when the documents reach the job office, the package or submittal number is assigned and the receipt date is logged under *Preparation and Receipt.* When the documents are sent to and then returned by the architect and/or owner, these dates are logged under *Review/Approval,* and so on.

The problems with record keeping arise primarily because the volume of documents becomes so great on many jobs. In such cases, hard and fast rules should be pre-established regarding the *processing* of documents. It is important, for example, to establish the maximum amount of time that the documents are permitted to stay in the contractor's offices being reviewed. This time limit varies from contractor to contractor, but should be consistent within the company to ensure that the contractor is not the cause of any delays. Equally important is the establishment of procedures for *transmitting* documents. The project manager should ensure that documents are not allowed to accumulate for several days after approval; there should be a requirement that all processed documents be sent out no later than the following day. Whatever the specific rules, they should ensure *regular, daily* handling of all submittal data.

Coordinating Submittals with the Construction Schedule

One of the keys to effective project management is knowing that *the procurement process should serve the construction schedule, not the other way around.* If work on the job site is to proceed efficiently and smoothly, then the material must reach the craftsmen before it is to be installed. Otherwise, the work sequence is interrupted and must be altered to accommodate disjointed delivery schedules. This situation cannot help but have an adverse effect on overall project time and profit.

To ensure timely delivery of materials, the project manager needs a way to determine from the construction schedule the latest acceptable delivery dates. He must then take all possible action to ensure delivery by that date. The best way to accomplish this goal is to treat the submittal data and procurement tasks as *activities,* just like the construction activities. Figure 7.6 shows that it is possible to convert the sequence of *prepare purchase order, prepare submittal data, review and approve submittal data,* etc. — as shown on the log — to a sequence of critical path activities which are (like construction activities) dependent on one another. The submittal and procurement *activities* can then be assigned scheduled durations in the same way, and the entire string of events which make up the procurement process can be tied to the appropriate construction activity.

For example, the activities necessary to purchase hollow metal door frames are shown in Figure 7.7. The times for these activities are based on various factors, as follows. *Prepare purchase order,* is an activity within the control of the project manager and can therefore be assigned a time according to job

Shop Drawing/Submittal Control Log

Spec. Section	Description	P.O. or Sub-contractor		Preparation for Submission	A/E Review & Approval		Fabricate & Deliver			Comments
		No.	Date Sent	Received From Sub.	Date to A/E	Date From A/E	Transmit to Sub.	Deliver to Site	Install	
				SCH						
				ACT						
				SCH						
				ACT						
				SCH						
				ACT						
				SCH						
				ACT						

Figure 7.2

Subcontractors Shop Drawings/Materials Data

Greene & Associates, Inc.

100 Smith Lane
Kingston, MA 02364
Tel: 555-1212

Project _____
Our Job No. _____

Contractor _____
Trade _____

Material/Equipment Item	Contract Award Date	Shop Drawing Submittal			Review			Fabricate & Deliver		
		Est. Weeks	Est. Date	Actual	Return Date	Status		Est. Weeks	Est. Date	Actual

Figure 7.3

Subcontractors Shop Drawings/Material Control

Greene and Associates
Construction Management Consultants

100 Smith Lane
Kingston, MA 02364
Telephone: 555-1212

Sheet ___ of ___
Date ___

Project ___
Our Job No. ___

Contractor ___
Trade ___

Spec. Sect.	Item	Contract And Date	Shop Drawing Submittals			Arch. Review/Approval			Fabrication & Delivery			Work Item Tie-In Point
			Sched. Submit Date	Crit. Submit Date	Actual Submit Date	Sched. Appr. Date	Crit. Appr. Date	Actual Appr. Date	Sched. Del. Date	Crit. Del. Date	Actual Del. Date	

Figure 7.4

procedures. Other activity times can be determined by talking to the parties responsible for them: the architect may say that *review and approval* will take four weeks; the subcontractor may require four weeks to *prepare shop drawings*, and 17 weeks to *fabricate and deliver* the hollow metal frames. For each time period indicated by each party, the project manager should add extra time to account for turnaround at the job site. In the case of hollow metal door frames, the adjusted times for each activity are as follows.

Prepare purchase order	5 days
Prepare shop drawings	22 days
Review and approve shop drawings	22 days
Fabricate and deliver door frames	85 days
Total time for submittal data, review, fabrications, & delivery	134 days

Figure 7.8 is a planning schedule showing the first activity that requires hollow metal door frames, Activity 50020, "Erect HM frames /studs— main lvl." The date provided for the early start of this activity is July 28, 1988. This date occurs 127 working days from the beginning of the job, which means

Greene & Associates
Construction Management Consultants

Shop Drawing Checklist   Project: _____
Date: _____   Project No.: _____

Specification Section	Submittal Item	Submitted to Architect?			Approved?			Returned to Subcontractor		
		Yes	No	Date	Yes	No	Date	Yes	No	Date

Figure 7.5

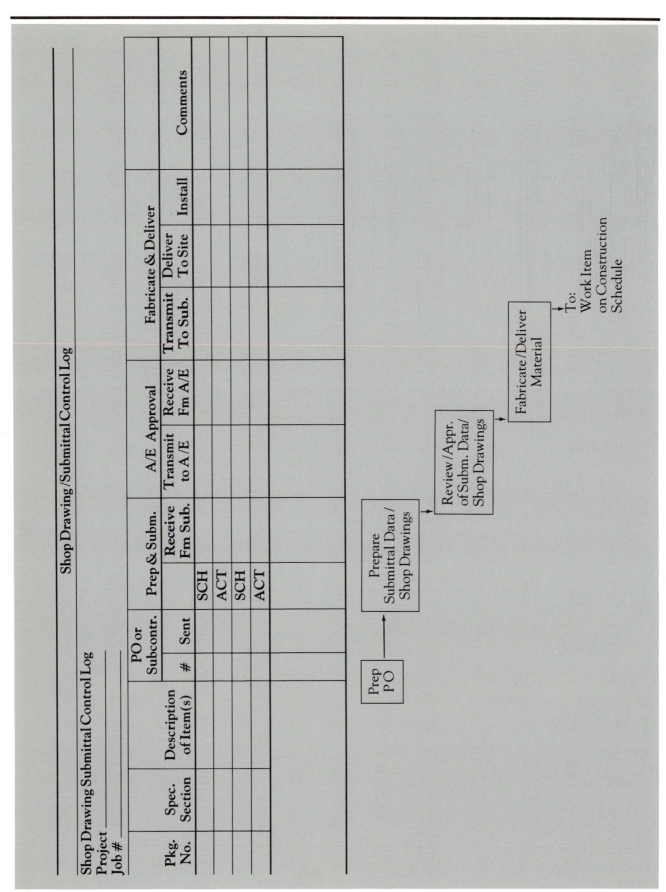

Figure 7.6

116

Project Planner

Activity ID	Orig. Dur.	Rem. Dur.	Pct.	Code	Activity Description	Early Start	Early Finish	Late Start	Late Finish	Total Float
1010	5	5	0	OFFC	Prepare Purch. Order — HM Door Frames	1FEB88	5FEB88	5OCT88	11OCT88	177
1020	22	22	0	OFFC	Prep. Subm. Data — HM Door Frames	8FEB88	8MAR88	12OCT88	10NOV88	177
1030	22	22	0	OFFC	Rvw. & Appr. Subm. Data — HM Door Frames	9MAR88	7APR88	11NOV88	12DEC88	177
1040	85	85	0	OFFC	Fab. & Del. — HM Door Frames	8APR88	4AUG88	13DEC88	10APR89	177

Figure 7.7

Project Planner

Activity ID	Orig. Dur.	Rem. Dur.	Pct.	Code	Activity Description	Early Start	Early Finish	Late Start	Late Finish	Total Float
50010	10	10	0	HVAC	Rough-in HVAC-Main Lvl	14JUL88	27JUL88	23DEC88	5JAN89	116
50020	10	10	0		Erect HM Frmes/Studs— Main Lvl	28JUL88	10AUG88	6JAN89	19JAN89	116
50040	10	10	0		Rough Plumbing— Main Lvl	11AUG88	24AUG88	20JAN89	2FEB89	116
50030	8	8	0	ELEC	Rough in Elec— Main Lvl	11AUG88	22AUG88	24JAN89	2FEB89	118
50050	6	6	0		Rough In Spnklr Piping, Main Lvl	11AUG88	18AUG88	26JAN89	2FEB89	120
50060	10	10	0		Hang & Tape GWB— Main Lvl	25AUG88	7SEP88	3FEB89	16FEB89	116

Figure 7.8

that, given the times provided to the project manager for the submittal and procurement tasks for hollow metal doors, the first doors will reach the site approximately nine days after the early start of the first activity for which they are needed.

Clearly, the project manager must take action to speed up some part of the submittal data/procurement process if he wishes to meet the early start schedule. There are several options at this point. One is to simply allow the installation to start a little later since it has over 100 days of float. If, however, there is no float, the situation is quite different. In this case, the project manager might have to make the architect aware of the need for a particularly quick review in this one case. Or, he could contact the supplier of hollow metal doors, and ask that special attention be given to the doors to be installed in the shear walls on the second floor. In any case, scheduling and then enforcing the time limits on submittal data tasks is just as necessary to the timely completion of the job as is the monitoring of construction activities. The key is maintaining a good tracking system.

Scheduling the Procurement Activities

Scheduling procurement activities is relatively easy from an administrative standpoint. The basic sequence of events is typically standard for a given project, and a project manager can set up a chain of activities for each procurement item, as shown in Figure 7.9. These chains of activities can then be connected to the construction logic diagram, using the hexagonal symbol to show the connection. The forward pass, backward pass, and float calculations work in exactly the same manner as before, and a schedule of start and finish times for procurement activities can be generated just as they are for construction.

The project manager must then decide whether or not the procurement should be connected to the construction activity as part of the overall schedule, and calculated at the same time. Based on the principle that the procurement schedule should not drive the construction schedule, the answer is that they should *not* be connected, nor calculated concurrently. In

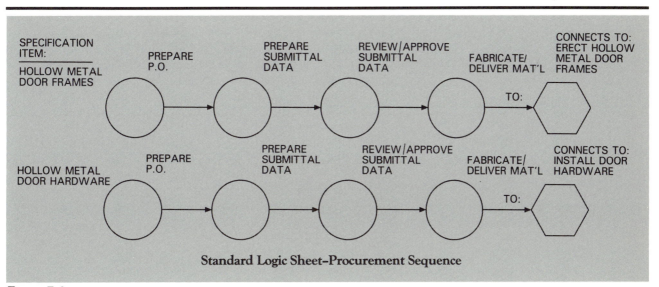

Standard Logic Sheet–Procurement Sequence

Figure 7.9

the case of hollow metal door frames, the early start of activity 50020 could be delayed a week to August 5, 1988. In this specific case, such an adjustment should not be a problem, since there is so much float available on the activity. The project manager might never realize that the procurement is determining the start of the erection of the hollow metal door frames. This lack of awareness is not ideal from a project management control standpoint. In a large project schedule where many procurement chains are connected to the construction schedule, it can become very difficult to determine where the procurement is affecting the overall project. Practically speaking, it is probably better to maintain a separate CPM schedule for the procurement activities, then transfer the dates to the procurement log and compare them by hand to the affected activities on the construction schedule. In this way, the project manager can be sure that the integrity of the construction sequence he has set up is maintained.

Reporting

Once the tracking procedures are underway and operating normally, the project manager must use the information developed to see that the submittals and purchases are carried out. The first step is to determine where the hold-ups exist in the process. The data must be viewed from two perspectives.

First, which pieces of submittal data are not meeting the scheduled dates? This determination should be done no less than weekly. It requires a detailed examination of the control log, and flagging of those submittals which have not met their dates. The list of flagged items should be broken down by the party responsible, i.e., subcontractors, architect, owner, etc. Separate lists should be made for each responsible party, showing how many submittals are in their hands, and the status of each submittal relative to the scheduled date. These lists can then be used in weekly job review meetings or in one-on-one sessions to review status and devise solutions for problem areas.

Developing status lists for submittal data can be done manually, by simply going through each log one at a time and picking out the behind-schedule submittals. This task is much easier if there is a separate log maintained for each subcontractor and supplier. The submittal control log can be kept on an ordinary microcomputer spreadsheet program, which will make the analysis somewhat less tedious. If a spreadsheet program is used, it is a good idea to keep a hand log as well, for back-up purposes.

Follow-up on the Information

Finally, follow-up is essential to the successful management of submittal data and procurement. The daily pressures of a construction project are such that it is very easy to fall into a pattern of performing the update and review on only an occasional basis. Unfortunately, this is an almost certain path to disaster. The procurement process requires regular, consistent, thorough record keeping and follow-up if the materials are to arrive on time, and if the process is to support the construction effort and not vice versa. In the case of our hollow metal door frames, 17 weeks is a long time for a purchase order to be out. This situation needs follow-up at no less than monthly, and probably two-week intervals. With this kind of regular monitoring, the manufacturer of the hollow metal door frames cannot let the fabrication schedule slip without the project manager knowing about it.

8

Organizing the Schedule

Organizing the Schedule

Scheduling using CPM techniques can involve a considerable amount of detail. Actually, detail should be kept to the minimum necessary to accomplish all aspects of the job. As projects grow larger, however, a high level of detail is inevitable. This chapter is devoted to techniques which make it possible to use a schedule efficiently even where significant detail is involved. The tasks required to organize and effectively use the schedule may seem too complicated to be worthwhile. In fact, quite the opposite is true. A moderate amount of organization up-front yields enormous dividends in the production of essential information.

Why Organize the Schedule?

There are two primary reasons for taking action to ensure that the schedule is designed and handled in an organized way. The first relates to efficient management of the actual construction process; the second is concerned with managing the schedule data and information — either by hand or machine.

The first reason for organizing the schedule involves efficient management of the construction process. Any project manager knows that there is simply never enough time to deal with every problem that arises. It is therefore imperative that project managers at all levels address only current issues, and that time not be spent on tasks that are not of immediate concern. The project manager cannot afford to be stricken with "information overload," as it hampers his ability to sort out the critical elements, or those that are in trouble and need immediate attention.

It should also be recognized that not all managers need the *same* information. For instance, a company vice president requires very different information from that needed by the project manager. Likewise, the superintendent requires a different set of data from that which is useful to the vice president and project manager. If the schedule is not set up to handle these differing needs, then the vice president is forced to wade through information intended for the project manager and superintendent, and vice versa.

To avoid confusion, the schedule must be organized in such a way that only the important and appropriate elements are presented to each party at the proper time. The manager must *not* be presented with information concerning portions of the job that are not currently relevant. For example, when the sample hotel is in the structural stage (i.e., the waffle slabs are being erected and placed), a project manager cannot afford to concern himself with future activities like roofing, or the fifth floor interiors schedules, nor activities that have already occurred, like the foundation schedules. While the structural work is proceeding on the waffle slabs, the project manager must zero in on that portion of the project, undistracted by other past or future activities. The project manager should not be presented with information that is intended for the vice president or others.

The second reason for organizing the schedule is to make the data manageable. During the course of the job, it is always necessary to update various activities, and to periodically change the sequence of construction and associated logic. When this occurs, it is helpful if project personnel who maintain the schedule are provided with organized, readily usable data. For example, suppose that during the erection of the waffle slabs it becomes necessary to change the logic. If it is difficult to locate the nodes associated with structural work on the several sheets of logic diagrams, then the administrative work on the schedule change becomes that much less efficient. In the worst case, the schedule is disorganized, and maintaining it becomes too burdensome. As a result, it may not be maintained at all by project personnel.

Levels of Detail

In organizing the schedule, level of detail is a very useful and important concept. The basic principle is as follows: the higher the manager, the lower the level of detail that is needed. Figure 8.1 illustrates this concept, showing three levels of management: vice president, project manager, and superintendent. Each of these managers has a different need for information, but all can be served by the same schedule provided that the schedule is properly organized.

First, consider the requirements of the superintendent. A working superintendent is interested in small, individual elements of the job. Again, we will use as an example the foundation of the sample hotel. Overall, the foundation consists of footings, columns, walls, and slab on grade. While the footing work is being carried out, the superintendent is not really interested in columns; these come later and will then be a major concern. For the present, his concern is with individual footings and the labor for the construction tasks that make up those footings, i.e., excavating each one, forming, reinforcing, placing, and stripping. Figure 8.1 is a schedule providing the planned starting times for each of these highly detailed tasks, all of which represent a real time concern to the superintendent.

Next, consider the project manager. His concern is with the individual elements of the foundations, since he must plan for reinforcing deliveries, purchase of formwork materials, subcontractor notification, etc. Note that the project manager is not concerned with the assignment of labor on individual days, only with the start of major elements of the overall foundation. Footings overall are the sum of all the lower level tasks (which are the responsibility of the superintendent). The project manager's requirements are, therefore, for a lower level of detail than that required by the superintendent.

Finally, consider the vice president. He is probably only interested in whether or not the overall foundation phase of the building is on schedule, and as such does not care if the individual footings, wall, or columns are on schedule. He certainly is not concerned with the individual placements of

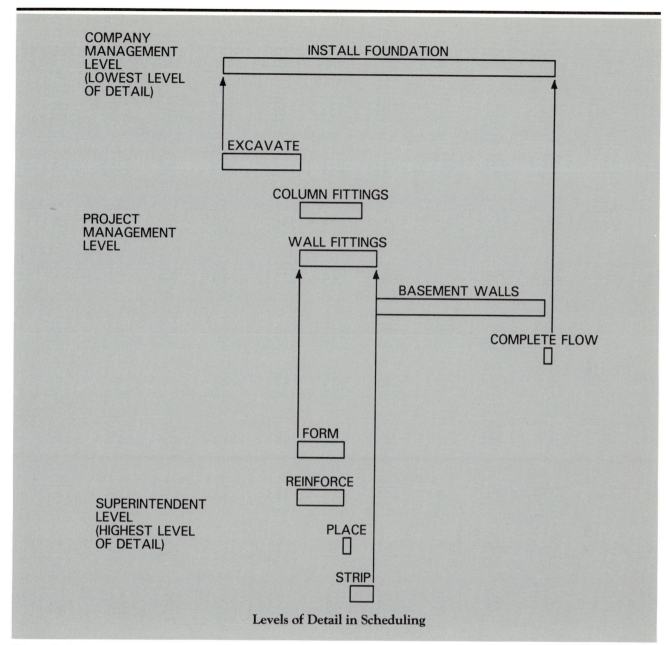

COMPANY
MANAGEMENT
LEVEL
(LOWEST LEVEL
OF DETAIL)

INSTALL FOUNDATION

EXCAVATE

PROJECT
MANAGEMENT
LEVEL

COLUMN FITTINGS

WALL FITTINGS

BASEMENT WALLS

COMPLETE FLOW

FORM

REINFORCE

SUPERINTENDENT
LEVEL
(HIGHEST LEVEL
OF DETAIL)

PLACE

STRIP

Levels of Detail in Scheduling

Figure 8.1

125

concrete and whether they are occurring on schedule. For one thing, the vice president is probably responsible for several other projects and can only spend so much time on each. The information presented to the vice president is at a much lower level of detail than that provided to those below him.

Tasks Required to Provide the Right Information

So far, we have described the concept and established the necessity for organizing the schedule. The question remains of how to do it. Specifically, the schedule must be organized in such a way that the three key tasks necessary to the development of good information can be accomplished. These tasks are as follows.

1. **Selection** —First, it is necessary to separate certain information from all other information. For example, if the superintendent wants information about the forming of footing line A alone, then our schedule system must allow us to identify each element of footing line A separate from all other work in the job. Or, the superintendent may wish to look only at forming of footings, and ignore placement, reinforcement, etc. Similarly, the schedule should be designed to permit that kind of selection.
2. **Sorting** — In addition to selecting, we must also be able to sort and present the data in an order that has some meaning to the people reviewing it. In the case of footing line A, the superintendent would probably want to list each element in the order in which it is supposed to occur.
3. **Summation** — It is necessary that the data can be summarized as shown in Figure 8.1. For example, if the project manager needs to check on the progress of the footings in general, but does not wish to see the detail, then the system must be capable of summarizing all footing activities into a single activity for display.

Finally, it is important to be able to accomplish all of these tasks at one time. For instance, if the vice president wants information about the foundation as a whole, and does not want to see information about the structure above or other elements, then the schedule must permit selection of foundation activities (excluding all others), and a summary of that set of activities. If the foundation is presented relative to other work, then sorting is also necessary.

This detailed process of selecting, sorting, and summarizing may at first glance seem too difficult and complicated to be worthwhile. Fortunately, this is not the case. All up-to-date computerized schedule systems are designed with these tasks in mind. The key to carrying out these tasks lies in the *coding* of activities, using the schemes provided by the systems. In reality, very little work is required to create a well organized, useful schedule.

Types of Coding Schemes

The key to organization lies in the coding scheme devised for the project. To effectively use a coding scheme, the scheduler must understand the following:

1. How to develop a coding scheme for a particular project
2. How the computer coding system works

To illustrate computerized schedule organization, two coding schemes are shown in the "Computer Coding" section later on in this chapter. One version is based on the *QWIKNET* scheduling system (produced by Project Software and Development, Inc., Cambridge, Mass.); the other is based on the *PRIMAVERA* system (Primavera Systems, Bala Cynwyd, PA.). Since these systems represent the two major approaches to coding used in computer systems today, both are included.

Coding Schemes for a Project

The coding systems shown for the sample hotel project are examples of types that allow the project manager to display project information in a number of different ways. These schemes are not the only ones that could be used, nor must a coding scheme be as complex as these in order to be useful or effective. They are intended only as representative types of schemes.

Coding by Project Phase

Most construction projects are carried out in a series of logical phases, with certain milestones marking the end of each phase. For example, a simple house might consist of the following phases:

1. Foundation
2. Framing and roof
3. Close in
4. Rough-in interior
5. Finish building exterior
6. Finish building interior
7. Landscaping/site work

Similarly, the sample hotel project might be broken down into these phases.

1. Foundation — excavation, footings, foundation walls.
2. Waffle slab structure — columns and waffle slabs for the main and second floors.
3. Tower structure — columns, shear walls, and flat slabs which make up the second floors through the roof structure in the hotel tower.
4. Kitchen/dining structure — masonry and steel structure over the main floor dining and banquet area.
5. Main interior — rough-ins and finishes for all areas on the main floor.
6. Tower interiors — rough-ins and finishes for all hotel rooms on the second through sixth floors.

Coding by Project Level

High-rise or multi-level projects benefit from a scheme that allows the project manager to review work taking place at different levels. A typical scheme for the sample project might be as follows.

Foundation level
Basement level
Main floor level
Second floor level, etc.

Using this coding scheme based on the building's levels, along with a scheme based on the project's phases, provides the project manager with a complete picture. Given the capabilities of most computerized systems, it is relatively easy to include both. If, however, a hand scheduling system is used, it may be best to choose only one coding scheme in order to cut down on the scheduler's manual work.

Coding by Trade

It is often helpful to be able to view the sequence of operations of a particular trade or crew, without regard to building level. For example, if the general contractor building the sample hotel elects to do all concrete using his company's employees, then it would be a great help to the project manager if he could selectively display the activities involving the contractor's own employees or crews, while ignoring activities involving subcontractors or other trades. This type of code is also helpful in identifying overlaps in scheduling crews, and for allocating labor.

Coding by Contractor or Subcontractor

In addition to reviewing activities involving one particular trade, it may also be helpful for the project manager to see the sequence of an individual subcontractor's work separate from all other work. In fact, subcontractors are usually eager for as much information as is available regarding the dates when they will be required to work, and they are apt to be more cooperative when they are better informed.

Other Coding Possibilities

While the above-mentioned coding schemes are generally useful on commercial building projects, there are several other types which are more appropriate for other kinds of work. For example, a contractor who does roadwork may wish to schedule and control bridgework separately from earthwork or paving. A contractor who builds large industrial plants may need to identify work on a horizontal or "geographic" basis. Home builders who construct large subdivisions might want to see or schedule work based on separate streets or by individual houses or groups of houses.

Computer Coding

In order to use coding schemes in a computerized CPM system, one must understand how coding works within typical scheduling systems. Most computerized scheduling packages on the market today use one of the following approaches.

1. Coding by groups
2. Coding by digit position

Coding Groups

The computer system for code groups is typically designed in such a way that the scheduler can represent the elements of his choice. To illustrate this type of coding system, samples of the hotel project are shown — using the PRIMAVERA system—in Figures 8.2 through 8.7. The first figure (8.2) shows an *activity codes dictionary*, listing terms defined in the scheduler. In this dictionary, the various code groups are listed with a short title and a long description. For example. RESP means *Contractor Responsible*. This code group is then broken down into a series of codes, which are assigned to the various contractors and subcontractors on the project. Note in Figure 8.3 that the activity code titles are ELEC for the electrical contractor, EXCV for the excavation contractor, and so forth. These codes can then be assigned to the appropriate activities. For example, *Mass Excavation* is assigned the EXCV code, and is thereby matched up with the responsible excavation contractor. The remaining codes for the sample hotel project are shown in Figures 8.4 through 8.7.

Activity Codes Dictionary Classification of Activity Codes			
	Name	Length	Description
1.	RESP	4	RESPONSIBILITY
2.	AREA	6	BUILDING PHASE
3.	SUMM	1	SUMMARY
4.	LEVL	4	BUILDING LEVEL
5.	TRAD	3	LABOR TRADE
6.		0	
7.		0	
8.		0	
9.		0	
10.		0	

Figure 8.2

Activity Codes for Project SAMP		
Code Name: RESP	**Description: Responsibility**	**Field Length = 4**
Code Value		**Code Title**
CRPT		Carpet Laying Subcontractor
ELEC		Electrical Subcontractor
EXCV		Excavation Subcontractor
GNCN		General Contractor
GWBS		Gypsum Wall Board Subcontractor
HVAC		Heating, Ventilating, and A/C Subcontractor
OFFC		Job Site Office Staff
PANT		Painting Subcontractor
PLMB		Plumbing Subcontractor

Figure 8.3

Activity Codes for Project SAMP		
Code Name: AREA	**Description: Building Phase**	**Field Length = 6**
Code Value		**Code Title**
EXTRSW		Exterior and Site Work Phase
FOUNDN		Foundation Phase
KDSTRU		Kitchen/Dining Structure Phase
MNINTR		Main Level Interior Phase
SUBDTA		Submittal Data Processing
TWRINT		Tower Interiors Phase
TWRSTR		Tower Structure Phase
WSSTRU		Waffle Slab Structure Phase

Figure 8.4

Activity Codes for Project SAMP		
Code Name: SUMM	**Description: Summary**	**Field Length = 1**
	Code Value	**Code Title**
	A	Foundation
	C	Waffle Slab Structure
	E	Tower Structure
	G	Dining Area Structure
	I	Basement Interior
	K	Main Level Interior
	M	Tower Interiors
	Z	Site Work

Figure 8.5

Activity Codes for Project SAMP		
Code Name: LEVL Description: Building Level Field Length = 4		
Code Value	Code Title	
2FLR	2nd Floor Level	
3FLR	3rd Floor Level	
4FLR	4th Floor Level	
5FLR	5th Floor Level	
6FLR	6th Floor Level	
BSMT	Basement Level	
FNDN	Foundation Level	
MAIN	Main Floor Level	
ROOF	Roof Level	

Figure 8.6

Activity Codes for Project SAMP	
Code Name: TRAD Description: Labor Trade Field Length = 3	
Code Value	Code Title
011	General Laborers
031	Formwork Carpenters
061	Finish Carpenters

Figure 8.7

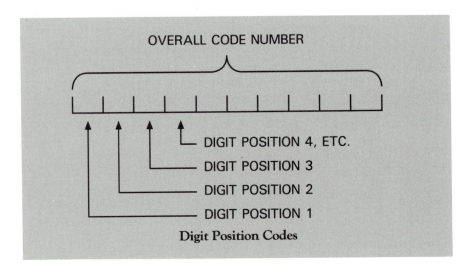

Figure 8.8

Coding by Digit Position

In this system, instead of providing code groups, one or two basic code *numbers* are provided. The codes are entered and located based on their position in the code number. Figure 8.8 illustrates the digit position concept, using samples from the QWIKNET computerized scheduling system, once again based on the hotel project. In this case, instead of assigning names or short letter groups to the various levels, phases, trades, etc., numbers are

Major Element	Code Structure — Sample Hotel		
	Qwiknet Digit Code	Primavera Group Code	Code for:
Phase of Building	1	FOUNDN	Foundation Work
	2	WSSTRU	Waffle Slab Structure
	3	TWRSTR	Tower Structure
	4	KDSTRU	Kitchen/Dining Structure
	5	MNINTR	Main Level Interior
	6	TWRINT	Tower Interiors
	7	EXTRSW	Exterior Work
	9	SUBDTA	Submittal Data
Level of Building	1	FNDN	Foundation
	2	BSMT	Basement
	3	MAIN	Main
	4	2FLR	2nd Floor
	5	3FLR	3rd Floor
	6	4FLR	4th Floor
	7	5FLR	5th Floor
	8	6FLR	6th Floor
	9	ROOF	Roof
Labor Trade or Type Work	1	031	Formwork Carpenters
	2	010	General Laborers
	3	061	Finish Carpenters
Responsibility	1	CARP	Carpet Laying Sub
	2	ELEC	Electrical Sub
	3	EXCV	Excavation Sub
	4	GNCN	General Contractor
	5	GWBS	Gypsum Wall Board Sub
	6	HVAC	Mechanical Sub
	7	OFFC	General Contractor (Office)
	8	PANT	Painting Sub
	9	PLMB	Plumbing Sub
Summary	1	A	Foundation
	2	C	Waffle Slab Structure
	3	E	Tower Structure
	4	G	Dining Area Structure
	5	I	Basement Interior
	6	K	Main Level Interiors
	7	M	Tower Interiors
	8	Z	Site Work

Figure 8.9

assigned which are inserted into the code's overall number. The overall numbering is shown in Figure 8.9, which also lists the equivalent codes in the PRIMAVERA grouping system. To code an activity, say *Form columns—main floor to second floor,* the overall code would appear as shown in Figure 8.10.

In this case, the number *2* in digit position one indicates *waffle slab structure;* the number *2* in digit position two means *waffle slab phase;* the number *3* in digit position three means *main floor level;* the number *031* in digit positions four through six means *formwork carpenters;* and the number *4* in digit position seven means *general contractor.*

Using the Coding Structures to Get Information

As noted earlier, the only purpose of the coding process is to make it easier for the project manager to obtain the information he needs — in an efficient, readable form — to run the construction job. To accomplish this goal, three tasks are necessary: selecting, sorting, and summarizing. The following figures illustrate this threefold process, using the sample hotel project.

Selection

Generally, selection is the first task involved in creating useful information from the schedule. Most importantly, logical sub-sets of activities should be selected to relate work by group. For example, Figure 8.11 is a schedule showing only foundation activities. This grouping is based on the AREA code for FOUNDN, which can be found in the *Codes* column. Another example is shown in Figure 8.12, a schedule for waffle slab work. This grouping process can be carried even further, as shown in Figure 8.13, in which waffle slab activities are selected for the main level only. In this case, the selection was based on the AREA code, *WSSTRU,* and the LEVL code, *MAIN.*

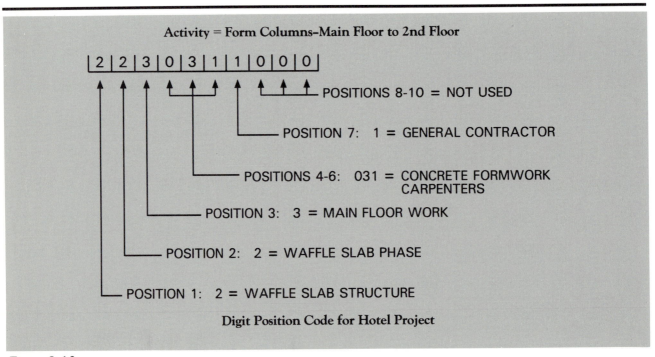

Activity = Form Columns–Main Floor to 2nd Floor

| 2 | 2 | 3 | 0 | 3 | 1 | 1 | 0 | 0 | 0 |

POSITIONS 8-10 = NOT USED

POSITION 7:   1 = GENERAL CONTRACTOR

POSITIONS 4-6:   031 = CONCRETE FORMWORK CARPENTERS

POSITION 3:   3 = MAIN FLOOR WORK

POSITION 2:   2 = WAFFLE SLAB PHASE

POSITION 1:   2 = WAFFLE SLAB STRUCTURE

Digit Position Code for Hotel Project

Figure 8.10

```
                              PRIMAVERA PROJECT PLANNER                                    SAMPLE HOTEL PROJECT

REPORT DATE 20MAR88  RUN NO.  31      SAMPLE HOTEL SCHEDULING PROJECT          START DATE 1FEB88  FIN DATE 7APR89

DAILY BARCHART SCHEDULE                                                        DATA DATE 1FEB88   PAGE NO.    1

                                                                                           DAILY-TIME PER.    1
```

.........ACTIVITY DESCRIPTION......... CODES FLOAT	SCHEDULE	01 FEB 88	08 FEB 88	15 FEB 88	22 FEB 88	29 FEB 88	07 MAR 88	14 MAR 88	21 MAR 88	28 MAR 88	04 APR 88	11 APR 88	18 APR 88	25 APR 88	02 MAY 88	09 MAY 88
ACTIVITY ID OD RD PCT																
MASS EXCAVATION 10010 7 FOUNDN 0	CURRENT	EEEEEE														
EXCAV TRENCH/WALL FOOTINGS 10020 1 FOUNDN 0	CURRENT	*	E													
REINFORCE WALL FOOTINGS 10040 1 FOUNDN 3	CURRENT	*	E													
EXCAVATE COLUMN FOOTINGS 10110 1 FOUNDN 28	CURRENT	*	E													
FORM WALL FOOTINGS 10030 4 FOUNDN 0	CURRENT	*		EEEE												
REINFORCE COLUMN FOOTINGS 10140 1 FOUNDN 30	CURRENT	*		E.												
FORM COLUMN FOOTINGS 10120 3 FOUNDN 28	CURRENT	*		EEE												
PLACE WALL FOOTINGS 10050 1 FOUNDN 0	CURRENT	*		E												
PLACE COLUMN FOOTINGS 10130 1 FOUNDN 28	CURRENT	*		E												
STRIP WALL FOOTINGS 10060 2 FOUNDN 0	CURRENT	*			EE											
STRIP FORMS COLUMN FOOTINGS 10150 2 FOUNDN 28	CURRENT	*			EE.											
REINFORCE BASEMENT WALLS 10080 2 FOUNDN 9	CURRENT	*			EE											
FORM BASEMENT WALLS 10070 11 FOUNDN 0	CURRENT	*			EEEEEEEEEE											
PLACE BASEMENT WALLS 10090 2 FOUNDN 0	CURRENT	*					EE									
STRIP FORMS BASEMENT WALLS 10100 11 FOUNDN 0	CURRENT	*						EEEEEEEEEE								
BACKFILL WALLS BASEMENT 10160 4 FOUNDN 0	CURRENT	*								EEEE						
COMPLETE FOUNDATION 19999 0 FOUNDN 0	CURRENT	*									E					

Figure 8.11

PRIMAVERA PROJECT PLANNER   SAMPLE HOTEL PROJECT

REPORT DATE 20MAR88 RUN NO. 32   START DATE 1FEB88   FIN DATE 7APR89
DAILY BARCHART SCHEDULE   SAMPLE HOTEL SCHEDULING PROJECT   DATA DATE 1FEB88   PAGE NO. 1
                                                                   DAILY-TIME PER. 1

| ....ACTIVITY DESCRIPTION........... | CODES | FLOAT | SCHEDULE | 04 APR 88 | 11 APR 88 | 18 APR 88 | 25 APR 88 | 02 MAY 88 | 09 MAY 88 | 16 MAY 88 | 23 MAY 88 | 30 MAY 88 | 06 JUN 88 | 13 JUN 88 | 20 JUN 88 | 27 JUN 88 | 04 JUL 88 |
ACTIVITY ID OD RD PCT																
FRPS W/S COLS,EAST, BSMT TO MAIN 20010 3	WSSTRU	0	CURRENT	EE													
FRPS W/S COLS-WEST, BSMT TO MAIN 20020 3	WSSTRU	13	CURRENT	.EEE													
FRP WFFL SLBS-NE, MAIN LVL 20030 8	WSSTRU	0	CURRENT	.EEEEEEEE													
FRPS COLS-NE, MAIN TO 2FLR 25010 3	WSSTRU	17	CURRENT			EEE											
FRP WFFLE SLBS-SE, MAIN LEVEL 20040 8	WSSTRU	0	CURRENT			EEEEEEEE											
FRPS COLS-SE, MAIN TO 2FLR 25020 3	WSSTRU	17	CURRENT					EEE.								
FRP WFFLE SLBS-SW, MAIN LVL 20050 6	WSSTRU	0	CURRENT					EEEEEE								
FRP WFFLE SLBS-NW, MAIN LVL 20060 6	WSSTRU	0	CURRENT						EEEEEE							
FRP WFFLE SLBS-NE, 2FLR 25030 8	WSSTRU	0	CURRENT							EEEEEEEE						
STRP WFFLE SLBS-SW, MAIN 20090 2	WSSTRU	22	CURRENT								EE.					
FRP WFFLE SLB-SE, 2FLR 25040 8	WSSTRU	0	CURRENT								EEEEEEEE.					
STRP WFFLE SLBS-NW, MAIN 20100 2	WSSTRU	16	CURRENT									EE				
STRP WFFLE SLB-NE, MAIN LVL 20070 2	WSSTRU	8	CURRENT											EE		
STRP WFFLE SLB-SE, MAIN LVL 20080 2	WSSTRU	0	CURRENT												EE	
STRP WFFLE SLB-SE, MAIN LVL 25050 2	WSSTRU	0	CURRENT												EE	
STRP WFFLE SLB-SE, 2FLR 25060 2	WSSTRU	0	CURRENT													EE
COMPLETE WAFFLE SLAB 25999 0	WSSTRU	0	CURRENT														E

Figure 8.12

```
REPORT DATE 20MAR88  RUN NO.  33        SAMPLE HOTEL SCHEDULING PROJECT      START DATE 1FEB88  FIN DATE 7APR89
DAILY BARCHART SCHEDULE                                                      DATA DATE  1FEB88  PAGE NO.       1
                                                                                    DAILY-TIME PER.  1
```

| ....ACTIVITY DESCRIPTION........ | | | | | | 04 APR 88 | 11 APR 88 | 18 APR 88 | 25 APR 88 | 02 MAY 88 | 09 MAY 88 | 16 MAY 88 | 23 MAY 88 | 30 MAY 88 | 06 JUN 88 | 13 JUN 88 | 20 JUN 88 | 27 JUN 88 | 04 JUL 88 |
ACTIVITY ID OD RD PCT	CODES	FLOAT	SCHEDULE																
FRP WFFL SLBS-NE, MAIN LVL 20030 8	WSSTRUMAIN	0	CURRENT			.EEEEEEEE													
FRPS COLS-NE, MAIN TO 2FLR 25010 3	WSSTRUMAIN	17	CURRENT				EEE											
FRP WFFLE SLBS-SE, MAIN LEVEL 20040 8	WSSTRUMAIN	0	CURRENT				EEEEEEEE											
FRPS COLS-SE, MAIN TO 2FLR 25020 3	WSSTRUMAIN	17	CURRENT					EEE.										
FRP WFFLE SLBS-SW, MAIN LVL 20050 6	WSSTRUMAIN	0	CURRENT					EEEEEE										
FRP WFFLE SLBS-NW, MAIN LVL 20060 6	WSSTRUMAIN	0	CURRENT						EEEEEE								
STRP WFFLE SLBS-SW, MAIN 20090 2	WSSTRUMAIN	22	CURRENT								EE						
STRP WFFLE SLBS-NW, MAIN 20100 2	WSSTRUMAIN	16	CURRENT									EE					
STRP WFFLE SLB-NE, MAIN LVL 20070 2	WSSTRUMAIN	8	CURRENT											EE			
STRP WFFLE SLB-SE, MAIN LVL 20080 2	WSSTRUMAIN	0	CURRENT												EE		
STRP WFFLE SLB-SE, MAIN LVL 25050 2	WSSTRUMAIN	0	CURRENT												EE		

Figure 8.13

The value of this selection process is, of course, that the project manager can concentrate on only the information that is pertinent at the time; all extraneous data can be excluded.

Sorting the Activities

As a practical matter, the task of sorting is simpler than selection, since only three sorts are of real value in most situations. The first is the *early start/early finish* (ES/EF) sort, showing work in the order in which it will normally occur. Figures 8.11–8.13 are sorted in this way, and clearly display the sequence of work as a superintendent or project manager would expect to perform it.

Also useful is the *Total Float/Early Start* (TF/ES) sort, shown in Figure 8.14. This schedule contains the same activities as Figure 8.11, but in this case, the critical path is obvious, since it is at the top of the page, and other activities are listed down the page as their float increases.

Finally, Figure 8.15 shows an *activity number* sort, which is useful when maintaining the schedule record. Projects having a large number of activities often make the schedule more difficult to maintain. A listing by their identification numbers helps in locating these items for updating and other purposes.

Summary of Activities

Lastly, Figure 8.16 shows a summary of all the activities in the sample hotel project, based on the SUMM code. The project is broken down into six basic phases. Each activity is assigned a SUMM code based on the phase in which it belongs. Looking at Figure 8.11, it can be seen that the first activity in the foundation starts at the beginning of February, and the last ends at the beginning of May. These times correspond to the overall FOUNDATION activity in Figure 8.16. The summarizing process provides a basis for presenting the current situation to visiting vice presidents or others who do not need detail, but rather overview information.

136

```
                          PRIMAVERA PROJECT PLANNER                          SAMPLE HOTEL PROJECT

REPORT DATE 20MAR88  RUN NO. 34      SAMPLE HOTEL SCHEDULING PROJECT          START DATE 1FEB88  FIN DATE 7APR89
DAILY BARCHART SCHEDULE                                                       DATA DATE  1FEB88  PAGE NO.    1
                                                                                          DAILY-TIME PER.   1
```

..........ACTIVITY DESCRIPTION.......... ACTIVITY ID OD RD PCT CODES	FLOAT	SCHEDULE	04 APR 88	11 APR 88	18 APR 88	25 APR 88	02 MAY 88	09 MAY 88	16 MAY 88	23 MAY 88	30 MAY 88	06 JUN 88	13 JUN 88	20 JUN 88	27 JUN 88	04 JUL 88
FRPS W/S COLS,EAST, BSMT TO MAIN 20010 3 WSSTRU	0	CURRENT	EE													
FRP WFFL SLBS-NE, MAIN LVL 20030 8 WSSTRU	0	CURRENT	.EEEEEEEE													
FRP WFFLE SLBS-SE, MAIN LEVEL 20040 8 WSSTRU	0	CURRENT			EEEEEEEE											
FRP WFFLE SLBS-SW, MAIN LVL 20050 6 WSSTRU	0	CURRENT					EEEEE									
FRP WFFLE SLBS-NW, MAIN LVL 20060 6 WSSTRU	0	CURRENT						EEEEE								
FRP WFFLE SLBS-NE, 2FLR 25030 8 WSSTRU	0	CURRENT							EEEEEEE							
FRP WFFLE SLB-SE, 2FLR 25040 8 WSSTRU	0	CURRENT								EEEEEEEE						
STRP WFFLE SLB-SE, MAIN LVL 20080 2 WSSTRU	0	CURRENT												EE		
STRP WFFLE SLB-SE, MAIN LVL 25050 2 WSSTRU	0	CURRENT												EE		
STRP WFFLE SLB-SE, 2FLR 25060 2 WSSTRU	0	CURRENT												EE		
COMPLETE WAFFLE SLAB 25999 0 WSSTRU	0	CURRENT													E	
STRP WFFLE SLB-NE, MAIN LVL 20070 2 WSSTRU	8	CURRENT											EE			
FRPS W/S COLS-WEST, BSMT TO MAIN 20020 3 WSSTRU	13	CURRENT	.EEE													
STRP WFFLE SLBS-NW, MAIN 20100 2 WSSTRU	16	CURRENT									EE					
FRPS COLS-NE, MAIN TO 2FLR 25010 3 WSSTRU	17	CURRENT			EEE											
FRPS COLS-SE, MAIN TO 2FLR 25020 3 WSSTRU	17	CURRENT				EEE										
STRP WFFLE SLBS-SW, MAIN 20090 2 WSSTRU	22	CURRENT								EE.						

Figure 8.14

SAMPLE HOTEL PROJECT

REPORT DATE 20MAR88 RUN NO. 38   SAMPLE HOTEL SCHEDULING PROJECT   START DATE 1FEB88   FIN DATE 7APR89

SCHEDULE REPORT - SORTED BY ACTIVITY NUMBER   DATA DATE 1FEB88   PAGE NO. 1

ACTIVITY ID	ORIG DUR	REM DUR	PCT	CODE	ACTIVITY DESCRIPTION	EARLY START	EARLY FINISH	LATE START	LATE FINISH	TOTAL FLOAT
20010	3	3	0	GNCNBSMT	FRPS W/S COLS,EAST, BSMT TO MAIN	31MAR88	4APR88	31MAR88	4APR88	0
20020	3	3	0	GNCNBSMT	FRPS W/S COLS-WEST, BSMT TO MAIN	5APR88	7APR88	22APR88	26APR88	13
20030	8	8	0	GNCNMAIN	FRP WFFL SLBS-NE, MAIN LVL	5APR88	14APR88	5APR88	14APR88	0
20040	8	8	0	GNCNMAIN	FRP WFFLE SLBS-SE, MAIN LEVEL	15APR88	26APR88	15APR88	26APR88	0
20050	6	6	0	GNCNMAIN	FRP WFFLE SLBS-SW, MAIN LVL	27APR88	4MAY88	27APR88	4MAY88	0
20060	6	6	0	GNCNMAIN	FRP WFFLE SLBS-NW, MAIN LVL	5MAY88	12MAY88	5MAY88	12MAY88	0
20070	2	2	0	GNCNMAIN	STRP WFFLE SLB-NE, MAIN LVL	8JUN88	9JUN88	20JUN88	21JUN88	8
20080	2	2	0	GNCNMAIN	STRP WFFLE SLBS-SE, MAIN LVL	20JUN88	21JUN88	20JUN88	21JUN88	0
20090	2	2	0	GNCNMAIN	STRP WFFLE SLBS-SW, MAIN	19MAY88	20MAY88	20JUN88	21JUN88	22
20100	2	2	0	GNCNMAIN	STRP WFFLE SLBS-NW, MAIN	27MAY88	30MAY88	20JUN88	21JUN88	16
25010	3	3	0	GNCNMAIN	FRPS COLS-NE, MAIN TO 2FLR	15APR88	19APR88	10MAY88	12MAY88	17
25020	3	3	0	GNCNMAIN	FRPS COLS-SE, MAIN TO 2FLR	27APR88	29APR88	20MAY88	24MAY88	17
25030	8	8	0	GNCN2FLR	FRP WFFLE SLBS-NE, 2FLR	13MAY88	24MAY88	13MAY88	24MAY88	0
25040	8	8	0	GNCN2FLR	FRP WFFLE SLB-SE, 2FLR	25MAY88	3JUN88	25MAY88	3JUN88	0
25050	2	2	0	GNCNMAIN	STRP WFFLE SLB-SE, MAIN LVL	20JUN88	21JUN88	20JUN88	21JUN88	0
25060	2	2	0	GNCN2FLR	STRP WFFLE SLB-SE, 2FLR	20JUN88	21JUN88	20JUN88	21JUN88	0
25999	0	0	0	GNCN	COMPLETE WAFFLE SLAB	22JUN88	22JUN88	22JUN88	22JUN88	0

Figure 8.15

```
PRIMAVERA PROJECT PLANNER                              SAMPLE HOTEL PROJECT

REPORT DATE 20MAR88 RUN NO.  37    SAMPLE HOTEL SCHEDULING PROJECT    START DATE 1FEB88  FIN DATE  7APR89

DAILY BARCHART SCHEDULE                                               DATA DATE  1FEB88  PAGE NO.   1

                                                                              WEEKLY-TIME PER.   1
.........ACTIVITY DESCRIPTION.........              07  04  02  06  04  01  05  03  07  05  02  06  06  03  01  05
ACTIVITY ID  OD  RD  PCT  CODES  FLOAT  SCHEDULE    MAR APR MAY JUN JUL AUG SEP OCT NOV DEC JAN FEB MAR APR MAY JUN
                                                    88  88  88  88  88  88  88  88  88  88  89  89  89  89  89  89

FOUNDATION                             CURRENT    EEEEEEEEEEEE.    .   .   .   .   .   .   .   .   .   .   .   .   .
                                                  *             *
WAFFLE SLAB STRUCTURE                  CURRENT    * *   EEEEEEEEEEEE    .   .   .   .   .   .   .   .   .   .   .
TOWER STRUCTURE                        CURRENT    * *     .     .   EEEEEEEEEEEEEEEEEEEEEEEEEE   .   .   .   .   .   .
DINING AREA STRUCTURE                  CURRENT    * *   EEEEEEEEEEEEEEE    .   .   .   .   .   .   .   .   .   .
MAIN LEVEL INTERIOR                    CURRENT    * *     .   .EEEEEEEEEEEEE   .   .   .   .   .   .   .   .   .
TOWER INTERIORS                        CURRENT    * *     .     .   .   .EEEEEEEEEEEEEEEEEEEEEEEEEE   .
```

Figure 8.16

9

Monitoring and Controlling the Project

Monitoring and Controlling the Project

While a good initial plan is essential, it is not enough to ensure a successful project. During the course of construction, uncontrollable events are bound to alter the original plan. The project manager must have a means of monitoring the effects of these outside factors. Once the deviations have been detected and measured, the project team must be mobilized to bring the project back on schedule. The monitoring process consists of the following steps.

1. **Monitoring Progress.** This step is frequently called *progress measurement,* or *updating the scheduling.* It is primarily a process of collecting detailed data on the work, then processing it in a computer or manual system to arrive at an accurate representation of the current job status. Monitoring progress corresponds with Steps 3 and 4 of the Project Control Cycle (see Chapter 2 for a definition and illustration of the Project Control Cycle).
2. **Comparing Progress to Goals.** The actual progress on the job is compared to the progress planned in the original schedule. This is Step 5 of the Project Control Cycle, and consists of displaying the data collected in the updating step. The project team uses this information to make decisions regarding future actions.
3. **Taking Corrective Action.** In this final stage, the project manager corrects any schedule problems based on all of the available information. Personnel and equipment are mobilized to carry out a new plan for finishing the job.

Communication

The controlling function depends on good communication at the construction site. A project manager who thoroughly monitors and compares, and then decides on a course of action to correct deficiencies, but does this without the participation of the subcontractors, superintendents, and other members of the project team, may find that his measures are doomed to failure. First, it is almost impossible to determine the status of the project without talking to the people who are actually carrying out the work. As good as they are, CPM systems have real limitations, and will not provide all of the details needed to get an accurate picture of where the job stands. CPM may provide a warning system indicating where a problem is occurring, but it does not diagnose the problem. Secondly, the project manager depends on the project team (subcontractors, superintendents, and others) to carry out whatever corrective action is needed. Members of the project team may not be highly motivated to solve problems unless they are involved in the decision-making process. In any case, if the team members are not competent to participate in the problem solving process, they probably should not be on the team in the first place.

Controlling is like the planning and scheduling processes that precede it in that the first vital step is to bring all pertinent parties into the process. Effective communication requires the following.

1. **Consulting personnel** during the monitoring process, since they are the best source of data regarding the individual parts of the job. This chapter covers the specific procedures for updating the schedule, but it must be recognized that the basic source of input is the project team.

2. **Displaying the information obtained in the clearest possible way.** This means using simple, straightforward, graphic displays which are understandable to everyone. As a general rule, the best means of communication are time-scaled bar charts developed from the CPM schedule. It is best to provide only the information that is pertinent to the tasks or activities being dealt with at that moment, thereby avoiding information overload. Fortunately, most of today's computer systems have the capacity to selectively organize and display information.

3. **Communicating regularly** with all parties in order to determine what corrective actions need to be taken. A project manager should hold regularly scheduled meetings that include all of the parties working on a job at that time. The following elements should be included in such meetings.

   • Attendance at the meeting should be required (preferably by contract) for all subcontractors and superintendents whose work is either under way or due to begin in the near future.

   • Every attendee should be given a photocopy of the schedule, showing progress during the last reporting period, and that which is anticipated for the next two reporting periods. For example, if the meetings were held weekly, the schedule would show a three-week time span. Provisions would also be made to project the schedule on an overhead or schedule board of some kind so that all parties can see the changes as they are made.

   • The meeting must be conducted in such a way that all parties can comment on what has happened thus far and what is being planned. By encouraging this kind of participation, an atmosphere is created which fosters public commitment to schedule performance. These kinds of commitments tend to be honored more faithfully than private ones, since reputation among one's peers can be a strong motivator. It is not necessarily critical to reach an absolute consensus on every decision, but it is important not to dictate to subs or superintendents without consultation.

   • The decisions that result from the meeting should immediately be published for all parties.

   • Finally, the meetings must be held without fail on a regular basis, and the decisions followed up by the project manager.

Monitoring Progress and Schedule Update

The goal of updating is to determine the present status of the job. In the simplest terms, is the job behind, on time, or ahead of schedule? It is also necessary to know which specific aspects of the job are behind. Thus, details of individual activities are also important. In order to perform updates, it is necessary to establish the start and finish times of activities, and if possible, their production rates.

In addition to recognizing currently active parts of the job, it is important to know how the present work status affects future work. This factor is known as the *downstream impact.* For example, it is possible to be approximately on schedule at a given time, but — if the correct details are not observed — to end up behind later on in the job — without realizing that the schedule is not going to be met. The capacity to measure such downstream effect is one of the most important aspects of the CPM technique, and serves as a very useful tool in planning recovery when behind schedule. It is also necessary to record the progress of the job for legal purposes. Good schedule records serve as the basis for claims, and as a defense against them.

Steps in Updating

In order to determine the present status of the job, a two-step process is necessary and must be taken in order. Bypassing one to get to the other leads to inaccuracy at best, and can be dangerous at worst. The two steps are:

1. Measure the progress of each activity individually. The updating need not occur in the order that the activities occur, especially since updating is often done sometime after the actual work is performed. It is important to update all activities before attempting to update the job as a whole.
2. Measure the impact of the activity progress on the job as a whole. This is done by using the information derived from the individual activities to recalculate the forward pass and backward pass, and then determining a new overall job duration.

How Often to Update?

The question always arises — how often should the monitoring process be carried out? The general practice in the industry is too infrequent. The value of a construction management system is in direct proportion to the timeliness of the information it provides. If the controlling process is not carried out often enough, or on a regular basis, the benefits are likely to be lost.

To be more specific, once a month should probably be the maximum period between updates and schedule meetings. Many construction jobs can be run with meetings held at monthly intervals, but in most cases, weekly meetings and updates are advisable. In fact, if the situation warrants, daily schedule updates may be necessary. An example of such a situation is where the potential for lost revenues from a facility being built is extremely high relative to the value of the construction on the facility. In any case, the benefits of updating by carrying out the control cycle must be measured against the costs. The benefits are difficult to measure precisely, but considering the costs of schedule delays, it is usually better to err on the side of too many schedule updates rather than too few.

Updating the Individual Activities

The status of individual activities can fall into several different categories, depending on how much progress has been made at the time the update occurs. Progress is reported on the basis of the concepts illustrated in Figure 9.1.

Each reporting cycle covers a predetermined time period; the status of each activity is reported from the perspective of a "time now" date, usually called a *data date.* As can be seen in Figure 9.1, the data date represents the end of the chosen reporting period. (In this case, a monthly reporting period is used.) The status of individual activities relative to the data date, or "time now" point, falls into one of the following cases, or a variation thereof.

Case 1 in Figure 9.1 is an activity which was started and completed prior to the data date. In this circumstance, the activity simply has an actual start (AS), and an actual finish (AF) reported.

Case 2 is more complex. The activity was started prior to the data date, but has not yet finished. The start is reported as an actual start (AS). The problem is determining present status and subsequent expected finish (ExF). It is important to be as accurate as possible so that the expected finish has the proper effect on the downstream impact calculations. The three methods for determining present status and expected finish are listed below.

1. **Percent Complete (PC)** This method involves determining how much of the total work is complete, calculating a rate of progress per day, and then extending this value to determine an expected finish. To establish a percent complete, a project team member must calculate — in the field — the percentage of work that has been accomplished as of the data date. The actual finish is subtracted from the data date in order to determine the number of days worked. The percent complete is then divided by the number of days in order to arrive at a rate of progress per day. The remainder of the work can then be divided by the rate per day to determine how many work days remain for this activity. This number is then added to the data date to determine the expected finish. This procedure is illustrated in the following formulas.

% per day = % done/days worked, or

% per day = % done/DD (Data Date) − AS (Actual Start)

then, DD + (% remaining/% per day) = ExF (Expected Finish)

Using a computerized scheduling system, the percent complete number is simply input and the calculations carried out automatically.

2. **Remaining Duration (RD)** Using this procedure, the number of remaining work days is estimated and this figure added to the data date in order to determine the expected finish. The following formula is used.

DD + RD = ExF

The estimate of the remaining duration can be anything from an educated guess to a carefully calculated figure based on field data.

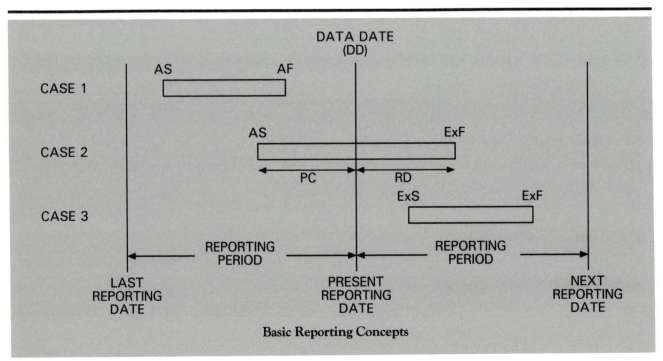

Basic Reporting Concepts

Figure 9.1

146

3. **Expected Finish (ExF)** For this procedure, the expected completion date is entered as a value. It is an appropriate method where the project manager has good reason to believe an activity will finish at a specific time, and wishes to display that fact in the update reports.

Problems with the Various Update Methods

All three of these methods have limitations and, therefore, should be used judiciously to ensure an accurate update. The problem with the use of actual starts and finishes is that this information is usually gathered long after the fact. Ideally, all job activity should be recorded on a daily job log. Unfortunately, entries are sometimes sporadic or incomplete. The person who reviews the reports at the end of the month often has to read between the lines to get a picture of the actual work done. Inadequate record keeping is a problem that also affects the percent complete method, since the actual starts may be missing as well.

The percent complete method tends to fall short primarily because detailed, accurate determinations are very seldom made for the actual work done. Usually, the percent complete is simply estimated, typically in 10% increments. As a result, the calculated expected finish may be even more inaccurate. Even when fairly accurate calculations are made, the production rate is often variable, or may be affected by stops and starts.

Finally, the expected finish method is often inaccurate because it depends on the prediction of a future event. This is especially difficult where both an expected start *and* an expected finish are involved. Also, the further the expected event is into the future, the more likely it is to be inaccurate.

So, the practical question remains — how to get the best possible information on the individual activities. The good judgment of the scheduler helps, but procedures can also be set up which will contribute to more accurate input and updates.

Daily Job Logs

First, it is vital that complete and accurate daily job logs be kept by all of the contractor's on-site personnel. This task is vital for reasons other than scheduling (e.g., claims) and the log can be made to serve more than one purpose. Samples of typical job log forms that collect good schedule information are shown in Figures 9.2, 9.3, and 9.4. The forms should be designed in such a way that the following information can be recorded.

1. Every schedule activity worked on that day should be identified. This record should include an activity description and number. Activities that begin or are completed that day should be noted.
2. The number of crews, workmen, and equipment involved in a particular activity should be noted. If available, their rates of production should also be recorded.
3. The areas of the project worked on by the crews should be identified. For example, if a crew is placing concrete in a building, the precise building elements placed should be recorded by floor and column line number.

If the job logs are kept accurately and faithfully, there is a reasonable chance that a scheduler (or whoever does the update) will have a good source of data for determining actual start and finish information.

Besides keeping good records, one must use the most effective updating methods possible. To summarize the three methods presented: Percent complete is probably the least reliable, and should be used only when the other methods cannot be used. Its potential inaccuracies are simply too great to yield valid data. Of the other methods, remaining duration and expected finish, both are only as accurate as the person providing the information.

Means Forms

DAILY
CONSTRUCTION REPORT

JOB NO.

DATE

PROJECT

SUBMITTED BY

ARCHITECT

WEATHER

TEMPERATURE   AM   PM

CODE NO.	WORK CLASSIFICATION	FOREMEN	MECHANICS	LABORERS	SUB-CONTRS	TOTAL HOURS	DESCRIPTION OF WORK
	General Conditions						
	Site Work: Demolition						
	Excavation & Dewatering						
	Caissons & Piling						
	Drainage & Utilities						
	Roads, Walks & Landscaping						
	Concrete: Formwork						
	Reinforcing						
	Placing						
	Precast						
	Masonry: Brickwork & Stonework						
	Block & Tile						
	Metals: Structural						
	Decks						
	Miscellaneous & Ornamental						
	Carpentry: Rough						
	Finish						
	Moisture Protection: Waterproofing						
	Insulation						
	Roofing & Siding						
	Doors & Windows						
	Glass & Glazing						
	Finishes: Lath, Plaster & Stucco						
	Drywall						
	Tile & Terrazzo						
	Acoustical Ceilings						
	Floor Covering						
	Painting & Wallcovering						
	Specialties						
	Equipment						
	Furnishings						
	Special Construction						
	Conveying Systems						
	Mechanical: Plumbing						
	HVAC						
	Electrical						

Figure 9.2

Means Forms

EQUIPMENT ON PROJECT	NUMBER	DESCRIPTION OF OPERATION	TOTAL HOURS

EQUIPMENT RENTAL - ITEM	TIME IN	TIME OUT	SUPPLIER	REMARKS

MATERIAL RECEIVED	QUANTITY	DELIVERY SLIP NO.	SUPPLIER	USE

CHANGE ORDERS, BACKCHARGES AND/OR EXTRA WORK

VERBAL DISCUSSIONS AND/OR INSTRUCTIONS

VISITORS TO SITE

JOB REQUIREMENTS

Figure 9.2 Cont.

Means Forms

PERCENTAGE COMPLETE ANALYSIS

PAGE

PROJECT

DATE

ARCHITECT

BY

FROM

TO

NO.	DESCRIPTION	ACTUAL OR ESTIMATED	TOTAL PROJECT	THIS PERIOD		PERCENT TOTAL TO DATE											
				QUANTITY	%	QUANTITY	10	20	30	40	50	60	70	80	90	100
		ACTUAL														
		ESTIMATED														
		ACTUAL														
		ESTIMATED														
		ACTUAL														
		ESTIMATED														
		ACTUAL														
		ESTIMATED														
		ACTUAL														
		ESTIMATED														
		ACTUAL														
		ESTIMATED														
		ACTUAL														
		ESTIMATED														
		ACTUAL														
		ESTIMATED														
		ACTUAL														
		ESTIMATED														
		ACTUAL														
		ESTIMATED														
		ACTUAL														
		ESTIMATED														
		ACTUAL														
		ESTIMATED														
		ACTUAL														
		ESTIMATED														
		ACTUAL														
		ESTIMATED														
		ACTUAL														
		ESTIMATED														
		ACTUAL														
		ESTIMATED														
		ACTUAL														
		ESTIMATED														
		ACTUAL														
		ESTIMATED														
		ACTUAL														
		ESTIMATED														
		ACTUAL														
		ESTIMATED														
		ACTUAL														
		ESTIMATED														
		ACTUAL														
		ESTIMATED														

Figure 9.3

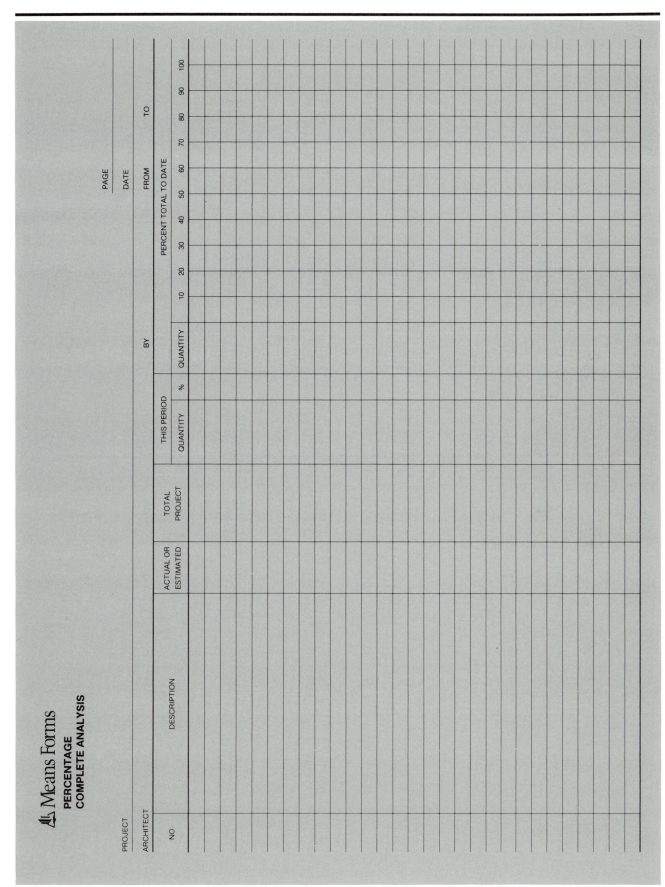

Figure 9.4

Both methods have the advantage that they create an atmosphere in which a sub or superintendent is asked to make a commitment to finishing an activity. As a result, better information may be available. It is important to realize that the individuals providing the information may be influenced by their own optimism or pessimism. Once again, this is a complication that the updater can overcome — by knowing his data sources, and using good judgment.

Other Areas to Check

In addition to the actual field activities, there are other influences that affect the schedule and these, too, must be checked. The following are examples.

1. **Changes in contractual dates.** Any change in contract dates, such as an extension of time, must be entered. These dates are typically recorded as constraints or "plugged dates."

2. **Changes in work sequences by field personnel.** It is not uncommon for field superintendents to perform work out of sequence without informing the project manager or scheduler of the changes. Work done out of sequence may not affect the schedule, but if the changes are major, they should be reflected in the update in the form of logic changes. This is particularly true if more work of the same type remains to be done.

3. **Changes in material delivery dates.** This is an area that can have a tremendous impact on projects, yet this information can be difficult to track. Ideally, a company should have some sort of a system for recording these changes. Unfortunately, this is often not the case. Gathering all the information about materials usually means consulting a variety of sources, such as telephone logs or submittal data logs, or interviewing purchasing agents or other individuals who are responsible for buying. At the very least, there should be a log of purchase orders and a correspondence file for each vendor on the job site. These files should contain all of the most recent information.

Measuring Progress on the Entire Job

Once the progress of all the individual activities has been determined, the progress of the entire job can be assessed. This task is relatively simple in concept, but can unfortunately be fairly cumbersome in practice.

To determine the progress of the job as a whole, information from the update of individual activities is used, and a new set of dates and times is calculated. This is done using the same calculation techniques shown in Chapter 6. The process of carrying out the forward and backward passes is no different at this point than it was for the original schedule, except that the actual and expected dates become fixed in both passes.

Because updating the entire job can be a cumbersome task, use of a computer CPM system offers a great advantage. A computer system is not foolproof, however, and care must be taken to avoid input errors. Also, it may be necessary to do several computer runs to ensure that the information is accurate before proceeding to the next step.

Comparing Progress to Goals

After the plan of action and the schedule for the project have been created, the project manager is then ready to carry out the ongoing task of monitoring the project and controlling the work as it proceeds. This established plan and schedule serves as the goal against which progress is measured in various ways. The attitude of the project manager toward time goals and performance is critical to maintaining the schedule. If he simply accepts schedule delays as inevitable and not correctable, then all the personnel on the project will do the same. If this attitude prevails, it is very likely that the delay situation will get even worse as the project proceeds. If, on the other hand, the project manager is persistent in conforming to the original plan of action and schedule goals, then the chances of staying on schedule are greatly improved.

Insisting on schedule compliance involves having a fixed goal that does not vary as circumstances change. The goal should be highly visible, and must always be held as a constant against which all work performance is measured. The goal is best established in the form of a *target schedule,* created at the beginning of the job.

Establishing the Target Schedule

Establishing the target schedule is a relatively simple task. The project manager does, however, have certain options in drawing up the target schedule, and making the correct choices is important to the long-run success of the project. The following paragraphs describe some of the approaches that the project manager can use to ensure the overall effectiveness of the scheduling process.

- The initial schedule contains both early and late times. It is generally best to set the target schedule based on early rather than late times, since this approach fosters an "earliest possible" perspective, and does not allow putting off an activity until the last possible start time. The idea is to promote an attitude that encourages getting on and off the job with each item of work as rapidly as possible.
- Some activities will be critical, while others will have float and can therefore be started at various times. It is best not to display or consider the float as part of the target schedule, since this approach defeats the sense of urgency created by the use of early start dates in the target schedule.
- It is important to note that there are sometimes legitimate reasons for changing the target schedule times. For example, the weather or delayed delivery of owner-furnished material may both be grounds for an extension of time for the project. If an extension is necessary, a new target schedule should be worked out and used. This new schedule must be clearly communicated to project personnel, and the reasons for the change explained thoroughly.
- The project manager may also decide to delay the start of certain activities in order to use manpower or equipment more efficiently. The target schedule should be changed to reflect this kind of modification. Once again, the change and the reasons behind it should be clearly explained to project personnel.
- Setting target schedules can be accomplished easily with most up-to-date computer CPM systems. The initial schedule or any updated versions can be designated as the target schedule. This plan can be stored and remain unaltered until the decision is made to change it. CPM systems are programmed to allow display of the target and the current schedule along side of one another, thereby making the comparison process much simpler.

Displaying the Results

Generally, schedules are most effectively displayed graphically, rather than in tabular or narrative form. Experience has shown that while bar charts are not the best means to do schedule planning or analysis, they are definitely the best means for showing schedule information. Most people can readily understand a standard bar chart format. Most computerized CPM systems on the market today provide the capability of creating these types of charts from the CPM logic and calculations.

Preventing Information Overload

It is important to direct the project's decision-makers to the major issues of a job at the time that these events are occurring. One of the computer's assets — the ability to store vast amounts of data — can actually present problems in that having this much information can make it difficult to zero in on the most critical issues. It therefore becomes important to abstract or sort out information for display, to highlight that which is pertinent, and ignore — for the moment — all other data.

To highlight the most pressing issues, the computer can be directed to display only *exception information.* Exception information pertains to situations that are not going according to plan. As long as there are only a few activities, it is not difficult for the manager to pick out the ones that are behind schedule. However, if there are 100 activities, then it is very important to have a method for singling out the late ones and flagging them for attention. The display created for the regularly scheduled meetings should clearly identify and separate the activities that need attention.

When displaying the schedule, only those parts that are currently being worked on should be shown. It serves no purpose, for example, to display schedule information pertaining to work completed three months ago, or work that will be performed four months in the future.

Finally, the experienced judgment of the scheduler or project manager plays a key role in deciding what information to display. It is important to remember that the schedule is only a tool, used to gather and display information that is needed to build the job. Rigid rules about how to run or display the schedule are completely inappropriate, and certainly unnecessary given the flexibility that modern CPM systems offer. It is often helpful to people in the field to see information in a familiar format. The schedule system should be used to accommodate personnel in this way, and not to force them into predetermined, unfamiliar procedures.

What to Look for in Project Reports

In reviewing update information, it is important to look beyond the job status as simply ahead or behind schedule. Other areas to be examined are:

1. What caused the job or parts of the job to fall behind schedule?
2. What events that have occurred thus far are likely to continue downstream as the job proceeds?

Answering the first question is important because knowing causes is essential to finding effective corrective actions. The answer to the second question is also key to the prevention of further delays. These questions are among the most typical. To get a complete picture of the job situation, the project manager and scheduler should look beyond the overall schedule results to these and other general points.

Status of Critical Activities   By definition, the critical path is the list of activities that must be started and completed on time in order for the job to be finished on time. This group of activities should immediately be examined to determine their status since failure to keep them on schedule will delay the project as a whole.

Non-critical Activities That Are Late in Starting   The problem with activities that start late is that they are more likely to finish late. This is true for reasons other than the obvious. The problem is that late-starting activities also tend to progress more slowly than they should. It is quite common to see activities that start before the late start date, which is acceptable, but then end up finishing beyond the late finish date. This puts them on the critical path. It is best to prevent this occurrence by detecting the problem at the start.

Activities with Low Production Rates   A related problem is activities that start on time, but are proceeding more slowly than planned. Early on, their progress may seem fine, but their duration may be extended, undetected, until an overly late finish.

Delays in Resource Delivery   This element is one of the most difficult to deal with in construction. As any experienced construction manager knows, material suppliers are forever moving delivery dates back. Much effort is spent trying to compensate for the delays. Fortunately, most delivery delays are grounds for extensions of contract time. Nevertheless, the fact of the delay must be recorded and its effects shown in the schedule. Most schedules use constrained dates to reflect material deliveries, or use a separate chain of activities for each major item. It is therefore good practice to check every constrained date or major material item on the project at the time of each update and control cycle.

Activities with More Downstream   Any activity that either starts late or shows a low rate of progress must be carefully watched. This is particularly true if there are more activities of the same type occurring later in the project. It may then be necessary to change the durations of the similar activities downstream, so that the schedule is realistic.

Changes in Outside Factors   Quite often, a scheduler concentrates on the field work, while ignoring the outside factors that also influence the schedule. For example, if a time extension has been granted, and the finish-no-later-than date for the project is not changed, then the calculated floats will be inaccurate. If the date has been moved back, the floats will be too large, and the field will be working with less flexibility than they actually have.

How to Find Out Why the Job Is Behind

Usually a scheduling system only informs the project manager which parts of a job are behind. It does not provide the reasons why. In order to truly know the condition of the job, the project manager must get out of the office and trailer and deal with the people in the field. While the schedule information points the way to where the problems lie, there is no substitute for face-to-face interviews with those responsible for the actual work.

The project manager must carry out the information-gathering interviews in a non-threatening manner. A superintendent or other supervisor who perceives that a project manager is out to "hang the guilty party" is more likely to provide self-serving information. As a result, the true causes of the job-site problems may go unnoticed, and the job continues in trouble. A far better approach is to encourage open communication and to involve the field supervisors in solving the problem. Most construction superintendents and foremen are quite proud of their achievements and want to do everything possible to maintain their reputations. This includes working toward an effective solution when problems arise.

Taking Corrective Action

The feedback loop is not complete until corrective action has been taken to bring the project back in line with the original plan. There are several points worth considering in this last part of the process.

The Necessity for Follow-up
The first point is that follow-up frequently does not occur. It is all too common for a construction company or project manager to set up a good

general plan for completing a job and then fail to properly monitor and control it. Effective follow-up requires a methodical and organized approach. Otherwise, it is all too easy to bypass them in the rush of daily concerns. Follow-up can be challenging in that it requires regular meetings, and many people in construction tend to regard meetings as an unproductive waste of time.

A positive attitude ("we will run this job, it is not going to run us") is also important. Too often, managers say that events inevitably come along which make a schedule useless, when what they should be saying is that a way will be found to meet the schedule, regardless of circumstances. This constructive approach is particularly important since those who choose to simply react to events almost guarantee failure on a construction job.

Types of Corrective Actions

It is impossible to cover all of the possible actions that a project manager can take to bring a schedule back in line. Nevertheless, there are some basic, sound measures which can be grouped into general categories, as follows.

Apply More Resources   The classic first response to schedule delay involves getting more people on the job, going to overtime, or mobilizing more equipment. Any of these responses may be appropriate under certain circumstances, but each must be considered carefully. For example, there is almost always an additional cost associated with speeding up work. The problem is that it is difficult to determine whether or not the costs of speeding up are less than the cost of a schedule overrun. It is clear that additional resources should only be applied to those activities that are on the critical path. Speeding up activities with float simply buys more float, which is of no value in completing the job earlier. Applying more resources can be a difficult decision. However, there are techniques available to assist the project manager in making these choices. These techniques are covered in Chapters 11 and 12.

Re-examine the Job Logic   In addition to applying more resources, the project manager also has the option of revising the sequence of activities on the job. For example, if the original job logic was done conservatively and used mainly finish-to-start lags, it is usually possible to go through the logic diagram and overlap activities more than was originally called for. Usually this is a better solution than applying additional resources. It does, however, call for closer control as the job progresses. It is, of course, possible to completely revise the sequence of operations to reflect an entirely new approach. This more drastic method would probably not be justified except in cases of extreme delay and penalties.

10
Maintaining the Schedule Records

Maintaining the Schedule Records

So far, we have concentrated on the running of a construction job. Attention must also be given to maintaining the schedule documents and the accompanying record keeping and paperwork. Having an established system for this aspect of the job reduces the burden and ultimately helps to make the schedule more effective by allowing managers more time to concentrate on construction rather than paperwork and computers. Good record keeping (particularly time records) is also crucial to the successful pursuit and defense of claims.

The text in this chapter is primarily concerned with the processes shown in Figures 10.1 and 10.2, which are schematics for the *Initial Development Phase* of a schedule, and for the *Reporting and Updating Cycle* which occurs each month throughout the life of the job. These procedures (which follow the *Project Control Cycle* arrangement) are by no means the only way to keep track of a schedule, but they are proven in practice and provide a suggested starting point from which a company can develop its own rules.

Initial Development Phase

Contract Award
The basic schedule cycle begins with the award of contract to the company, even though some preliminary schedule work may have been done for cost plus, negotiated, or construction management types of contract arrangements.

Making Activity Lists
The best tool for creating activity lists is probably a legal pad. It is often helpful to simply "brainstorm" a list and then organize it later into logical groupings. A computer can be useful for organizing such a list based on a set of criteria that are input. When using a computer for this purpose, the sequential breakdown approach described in Chapter 5 is most suitable.

The activity list should ultimately look something like the example shown in Figure 10.3.

Establishing the Work Sequence
In drawing logic diagrams, it is important to remember that the development of work sequences is very much a trial and error process. It is not just a matter of drawing, it is a process of thinking — using the logic diagram as a means of expressing those thoughts. Consequently, it is often best to start out with a

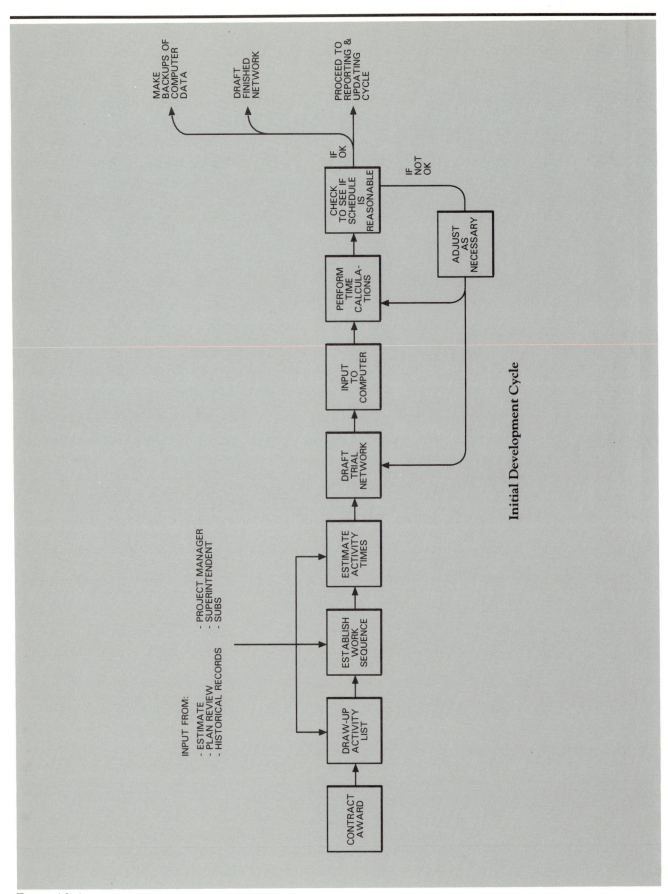

Initial Development Cycle

Figure 10.1

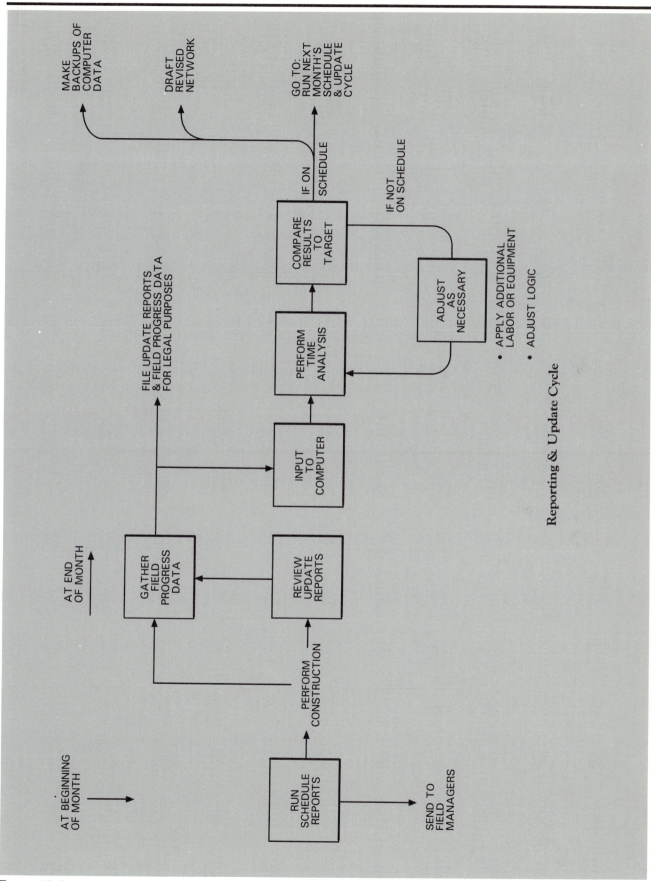

Figure 10.2

Reporting & Update Cycle

MAKE
BACKUPS OF
COMPUTER
DATA

DRAFT
REVISED
NETWORK

GO TO:
RUN NEXT
MONTH'S
SCHEDULE
& UPDATE
CYCLE

IF ON
SCHEDULE

IF NOT
ON SCHEDULE

COMPARE
RESULTS
TO
TARGET

PERFORM
TIME
ANALYSIS

ADJUST
AS
NECESSARY

• APPLY ADDITIONAL
  LABOR OR EQUIPMENT

• ADJUST LOGIC

FILE UPDATE REPORTS
& FIELD PROGRESS DATA
FOR LEGAL PURPOSES

INPUT
TO
COMPUTER

GATHER
FIELD
PROGRESS
DATA

REVIEW
UPDATE
REPORTS

AT END
OF MONTH

PERFORM
CONSTRUCTION

RUN
SCHEDULE
REPORTS

AT BEGINNING
OF MONTH

SEND TO
FIELD
MANAGERS

Activity List

I. Foundation
  A. Wall Footings
  B. Basement Walls
  C. Column Footings
II. Waffle Slab Structure
  A. Structure
    1. Columns — Basement to Main Floor
    2. Waffle Slab over Parking Garage
  B. Parking Garage
    1. Finish out Parking Garage
      a. Rough-in Electrical
      b. Painting
      c. Parking Lot Equipment
  C. Main Floor
    1. Envelope
      a. Masonry Wall
      b. Windows
    2. Interior
      a. Rough-in
        1) HVAC Ductwork and Equipment
        1) Frame Stud Walls
        3) Rough Electrical
        4) Rough Plumbing
        5) GWB
        6) Rough Paint
      b. Finish
        1) All Areas
          (a) Ceiling Grid
          (b) Vinyl Wall Covering
          (c) Light Fixtures
          (d) HVAC Registers
          (e) Doors and Hardware
          (f) Vinyl Asbestos Tile
          (g) Carpet
          (h) Finish Electrical
        2) Baths
          (a) Ceramic Tile
          (b) Finish Plumbing
          (c) Toilet Partitions
          (d) Bathroom Accessories
        3) Other
          (a) Casework
          (b) Storefront at Gift Shop
III. Tower (Note: These activities occur for each of the five floors.)
  A. Structure
    1. Shear Walls
    2. Columns
    3. Flat Slab Floors
    4. Masonry Walls @ Stairs and Elevator

Figure 10.3

162

B. Envelope
    1. Curtain Wall
    2. Windows
    3. Roofing and Sheet Metal at Top
  C. Interior
    1. Rough-in
      a. Frame Stud Walls
      b. Rough Electrical
      c. Rough Plumbing
      d. GWB
      e. Rough Paint
    2. Finishes
      a. All Areas
        1) Vinyl Wall Covering
        2) Light Fixtures
        3) Doors and Hardware
        4) Carpet
        5) Finish Electrical
        6) Install Furniture in Rooms
        7) Install In-Room Air Cond. Units
      b. Baths
        1) Ceramic Tile
        2) Finish Plumbing
        3) Toilet Partitions
        4) Bathroom Accessories
IV. Dining/Banquet Area
  A. Foundation
    1. Excavation
    2. Footings
  B. Structure
    1. Structural Steel
    2. Steel Joists
    3. Steel Roof Decking
  C. Envelope
    1. Masonry Walls
    2. Wall Insulation
    3. Windows
  D. Roof
    1. Roof Insulation
    2. Roofing and Sheet Metal
    3. Rooftop Mechanical Equipment
  E. Interior
    1. Rough-in
      a. All Areas
        1) HVAC Ductwork and Equipment
        2) Frame Stud Walls
        3) Rough Electrical
        4) Rough Plumbing
        5) GWB
        6) Rough Paint

Figure 10.3 Cont.

b. Kitchen
   1) Install Kitchen Equipment
   2) Casework and Cabinets
   3) Quarry Tile
2. Finish
   a. All Areas
      1) Ceiling Grid/Ceiling Tiles
      2) Vinyl Wall Covering
      3) Light Fixtures
      4) HVAC Registers
      5) Doors and Hardware
      6) Carpet
      7) Finish Electrical
      8) Folding Door in Banquet Room

V. Work Outside Building
   A. Porte Cochere
      1. Structure
      2. Facade
      3. Roofing
   B. Parking Lot
      1. Underground Electrical
      2. Underground Sprinkler Piping
      3. Curbs and Gutter
      4. AC Paving
      5. Street Lamps
      6. Sprinkler System
      7. Landscaping
      8. Striping

Figure 10.3 Cont.

pencil, eraser, and rough tracing paper, then progress through stages of refinement. Each succeeding stage will more closely approximate the best sequence for the job.

Ultimately, the logic diagram must be put into a finished form. There is no one format that must be followed. Any format is acceptable so long as it is clear and readable. Some examples of possible formats are shown in Figure 10.4.

While many computer programs have the ability to print logic diagrams, none are the equal of a good draftsman when it comes to displaying a large, complex diagram. There is really no substitute for a good paper and pencil logic development and record keeping process. An example of the type of logic diagram that must be maintained is shown in Appendix A.

In choosing a format for recording logic diagrams, it should be remembered that they will be altered frequently. One possibility is to draw the first diagram on a durable surface such as Mylar, and make blue line copies just before updates to logic are performed. The blue line can be kept as the recor copy, and only those changed portions of the Mylar original will need to be re-drawn.

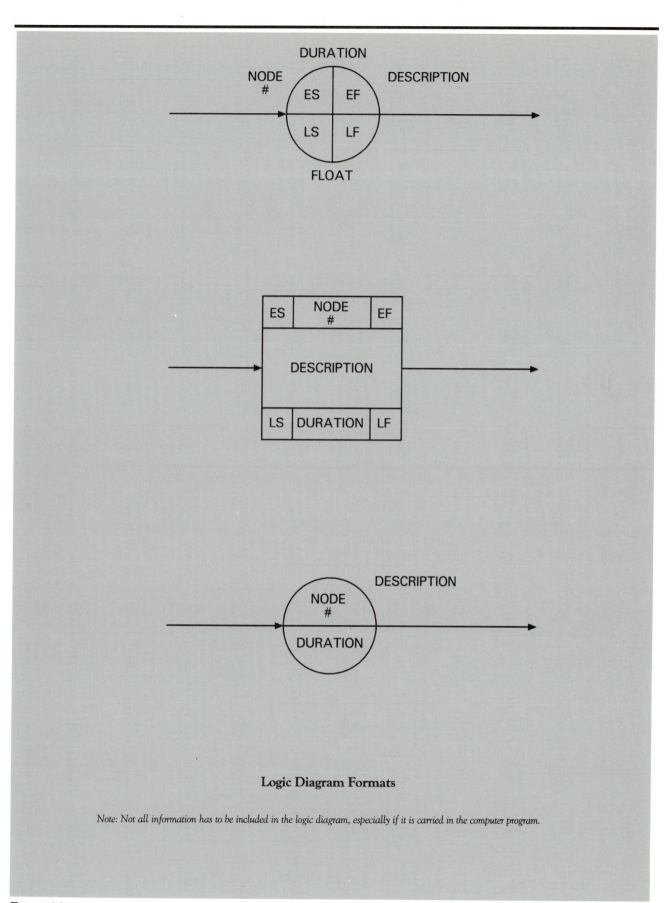

Logic Diagram Formats

Note: Not all information has to be included in the logic diagram, especially if it is carried in the computer program.

Figure 10.4

Estimating Activity Times

This part of the initial development can be complicated because there are so many different activities for which times must be assigned. Different activities are commonly mixed up, especially when the work is similar. It is therefore very important to record the productivity information in an orderly manner. In this way, confusion and inaccuracies are reduced as much as possible.

A standard activity analysis sheet is a good tool for organizing activities. An example is shown in Figure 10.5. The collection of sheets is kept in a looseleaf notebook, organized according to a practical classification system. The best classification system for scheduling buildings is probably the CSI MASTERFORMAT. (Information on this system is readily available from the Construction Specifications Institute. *Broadscope* headings are listed in

Activity Analysis Sheet

CSI Number: _____
Estimate Page Number: _____
Date: _____
Revision Number: _____

Activity Definition:
    Action: _____
    Object: _____
    Location: _____
General Description: _____
_____

Codes Involved:
    Cost code: _____
    Schedule Code: _____
Limiting Factors: _____
_____

Total Units of Work: _____
Productivity: MH/Unit: _____   EH/Unit: _____
Calculations:

$$\frac{MH}{Units} \times \text{\_\_\_\_Total Units} = \text{\_\_\_\_Total MH}$$

$$\frac{\text{\_\_\_\_Total MH}}{MH/Crew Day} = \text{\_\_\_\_Total Days}$$

Equipment Used: Type: _____   Qty: _____
_____   _____

Crew Description: _____
_____
_____
_____

Comments and Assumptions: _____
_____
_____
_____
_____

Figure 10.5

166

Appendix C.) Much of the productivity and cost data that will go into the time estimates is already organized in this format in the contract documents. It is also helpful if the activity analysis is cross-referenced by activity number, though this is not essential.

When maintaining the activity analysis records, it is also good practice to record the time and date of any and all updates. Since it is very common for a schedule to go through several evolutions, it is helpful to record the various phases, thereby avoiding time wasted in "re-inventing the wheel." It is also helpful if the individual schedule evolutions are numbered sequentially. For example, the first try, which should list all normal and standard construction conditions, could be Version Number 1.

Subsequent tries which include changes in strategy and construction procedures could be Versions No. 2, 3, etc. The changes made each time should be noted and recorded.

Inputting Data to the Computer

The logical approach to the task of inputting data is to avoid wasting the scheduler's valuable professional time. This is accomplished by setting up a system in which data can be input by trainees or other less experienced persons. The following paragraphs provide some general guidelines.

First, do not attempt to input data until the logic is "tied down" and the sequence of construction is decided. Entering data prematurely not only tends to be less efficient, but it is also very easy to lose track along the way of what has already been entered — and this leads to errors. A better approach is to first develop an activity list, and then a precedence list based on the completed diagram. Probably the easiest way to develop these lists is to have one person read off each item from the diagram, while another writes down the data in a tabular form such as shown in Figure 10.6. As the data is read off, it should be checked off on the diagram. This system prevents duplication of data and subsequent errors in input.

Second, once the lists are created, they can be used by less experienced persons to do the actual inputting of data. There is no one standard method for inputting data into computer systems due to the enormous variation in the systems available on the market today. It is worthwhile, however, to explore the available options. For example, programs that use "windows" (such as QWIKNET) work faster if all of the activities are input first, and then the precedences are input. The reason for this order is that window changes are slower than activity changes on the screen. The *PRIMAVERA* program, on the other hand, is probably faster if the activities and precedences are input at the same time, since the screen presents all the pertinent information at one time. Whatever system is used, the scheduler should explore the operating manual for hints on data input.

Third, after all data has been input, one should always print out a listing of the data, and compare it to the logic diagram one piece of information at a time. It is very rare to succeed in getting all the data input with precise accuracy on the first pass. Some data cleanup is almost always necessary.

Running the Trial Calculations

After all activity and precedence information has been input to the system, the network should be calculated. This means performing the forward and backward pass, and printing out a first run of the results.

The first results should be reviewed in a general way first, then in detail, if necessary. At this point, the scheduler is determining whether the overall time is reasonable and that it meets contract requirements. The overall flow of work should also be examined. The critical path should be checked to see if

10010	7	mass Excavation
10020	1	Excav. Trench/ Wall Footings
10030	4	Form Wall Footings
10040	1	Reinforce Wall Footings
10050	1	Place Wall Footings
10060	2	Strip Wall Footings
10070	11	Form Basement Walls
10080	2	Reinforce Basement Walls
10090	2	Place Basement Walls
10100	11	Strip Forms Basement Walls
10110	1	Excavate Column Footings
10120	3	Form Column Footings
10130	1	Place Column Footings
10140	1	Reinforce Column Footings
10150	2	Strip Forms Column Footings
10160	4	Backfill Walls Basement
19999	0	Complete Foundation

Figure 10.6

PRECEDENCE LIST

Predecessor	Successor	Lag
10010	10020	FS0
10020	10030	
10020	10040	
10020	10110	
10030	10050	
10040	10050	
10050	10060	
10060	10070	
10060	10080	
10070	10090	
10080	10090	
10090	10100	
10110	10120	
10110	10140	
10120	10130	
10140	10130	
10130	10150	
10150	19999	
10100	19999	

Figure 10.6 Cont.

169

it accurately reflects the opinions of field personnel as to which parts of the job are most important. If the job plan does not meet contract time, or does not appear to be reasonable and practical, then revisions must be made.

In the review process, some of the problems that are found are likely to be the result of inputting errors. This is a surprisingly common situation, even after several trial runs. Such errors must be completely corrected before any attempt is made to work out actual scheduling problems with project management personnel. Taking a schedule to people in the field and asking them to work on solutions based on inaccurate basic data compromises both the scheduler's credibility and the scheduling process. It also leads to inaccurate results.

When inputting errors have been corrected, it is time to address problems in the plan of action. The scheduler should approach field personnel with an attitude of, "we have a problem here, can you folks help?" Under no circumstances should a scheduler approach them with an attitude that assumes that the field plan was all wrong. After all, it is fundamental to the success of a schedule that it be treated as a tool to help management personnel, and cooperation is always better than confrontation.

Schedule Documentation

When putting together the initial schedule, it is important to always keep records of the changes and evolutions that occur. If a computer is used, it is easy to create back-ups of each run, thereby making a record of each new version. If the work is being done by hand, it is easy to draw the rough logic diagrams on reproducible paper, and then make prints for the purpose of keeping records at various stages. Rough copies of activity analysis sheets, notes, etc., should also be kept, preferably in an organized format.

Finally, it is vital to make a complete back-up, or copy, of the network that is to be used for construction. At later stages, this network will probably have to be altered. Nevertheless, it is a good idea to keep a copy of the original in a safe place.

Providing Information for Construction

Once the initial schedule has been established, and all parties to the job have agreed that it is feasible, the problem then arises of how to pass that information on to the users. It is particularly important that this data be accurately conveyed not only at the beginning of the job, but also at each interval. The schedule reports are primary tools for the ongoing scheduling of subcontractors and crews, and they must be effectively communicated.

Generally, it is best to establish a standard set of reports. More information can be added when appropriate, or when required by field personnel. The format chosen for a standard set of reports may be governed by the individual preference of the project manager. The following recommendations may offer a good starting point.

- Bar charts are generally superior to planning schedules for use as reporting devices since they provide the clearest display of schedule information. Thus, the basic working reports for the field should be daily interval bar charts, sorted according to early start/early finish. In this way, they clearly represent to management personnel the flow and sequence of work.
- Planning schedules (tabular format) should be provided as a back-up to the bar charts, since printed bar charts can sometimes be difficult to read precisely. In addition to the early start/early finish sort, a planning schedule should also be provided which is sorted by total float/early

finish. This sort format will show the critical path for the job at the top, so those crucial activities can be examined separately.

- Only the schedule information pertinent to the present time and work should be provided. It may be useful, for example, to single out only those activities that relate to a particular phase or level of the project, and ignore all others. A case in point is a building in the foundation stages. A bar chart showing only foundation activities makes it easier for managers to concentrate on the tasks at hand. Or, if interior work is being done, one might show bar charts for only the level that is currently being discussed. Confusion is reduced by minimizing the amount of extraneous information.

- "Windowing" should be considered if the system in use permits it. This technique allows for the specific selection of only those activities that are within a certain time frame, ignoring all others. A particularly good way to report scheduled work is to show the activity from the last reporting period, along with that which will take place within the next two reporting periods. For example, if a schedule meeting is held at the end of October, then the schedule report would show work from October 1st to December 31st.

- Use summary schedules for upper levels of management, and detailed schedules for field personnel. This approach upholds the principle of reporting appropriate information so as to not overload the managers. The superintendent must schedule subs and crews daily, whereas the vice president really only cares whether or not the job as a whole is on schedule.

- If using a computer, be creative in utilizing its capabilities to best advantage. Most systems available today permit some flexibility in generating reports, and the smart scheduler uses this computer power to best help the job. As a general rule, most computer systems can perform the same critical path computations; the real differences between systems lie in reporting capabilities.

- Finally, it is not always sufficient to simply distribute bar charts and assume that the field users will be able to interpret them. It is helpful to include a short one-to-two page memorandum or narrative report with the schedule reports to explain the problem areas. Further, it is often a good idea to discuss the situation directly with those who are involved, and be sure that they understand.

Reporting and Updating the Schedule

At the end of each reporting cycle, data must be collected in the field concerning progress to date. This information is entered into the system to carry out the monitoring function. After the data is recorded and the managers have fully reviewed the results and made new plans, the schedule is revised and new reports and information provided. To carry out this process as efficiently as possible, a regular procedure should be set up and performed without fail each reporting period. Following are some basic guidelines.

1. Just before the end of the reporting cycle, preparations can be made for gathering information. Most computerized systems have provisions for generating an *update report*. The report format typically provides spaces for entering all pertinent data, such as data date, actual start, percent complete, etc. This type of report is not always the most useful basis for discussing the actual happenings in the field. It is useful for recording information that will later be punched into the computer system.

   The most helpful tool for discussing information in the field is the bar chart, originally provided as a plan of action. If bar charts are marked with actual starts and finishes as the work proceeds, a clear, easy-to-read record is created. Even if the field managers do not mark the charts regularly, the scheduler can rerun the bar charts, take them to the field, and mark them while discussing the history of the last period with the . field managers.

A copy of the logic diagrams should also be made so that they can be reviewed for changes which have occurred. Sequences of activities commonly change in light of unanticipated events, and future computer runs should reflect the actual logic used. It is also very important that the historical record reflect actual events for claims purposes.

2. Once the reporting materials have been prepared, they can be taken out into the field and the information gathered. In addition to extensively interviewing field managers, it is also important to thoroughly review all daily job logs and such other records which carry histories of actual events.

3. When the data is gathered, it is entered into the system as *progress data*, and trial runs are made to see where the job stands relative to the original target schedule. At this point, the managers and schedulers can review the results and start to deal with the problems which have appeared in the course of the analysis. It is important to note here again that data input must be thoroughly checked for accuracy.

If new versions of the schedule are produced and proposed, it is a good idea to label the various versions according to their chronological order. For example, if several trial schedule changes are reviewed at the end of a project's fifth month, it might be advisable to label each of these trials as Version 5.1, 5.2, etc. Records should be kept of each change. The new schedule that is finally selected would be noted as Version 6.0, since it would be the plan for construction as of the beginning of month 6 of the job.

4. Finally, after the new revised plan of action is decided upon, a new set of reports is generated and sent to the field. It is vitally important at this point to make back-ups of the new plan — whether computer disk or hard copy — and to store them in a safe place separate from the working copy. It is also important that the records accurately reflect the progress to this point.

11

Resource Management

Resource
Management

One of the project manager's most important jobs is ensuring that all the resources necessary to build the project are available at the right time and place. While submittal data and procurement for material are very important aspects of project management, they are certainly not the only issues. Adequate labor of the right type or trade must also be available, as well as proper equipment for the trades to perform their work. Ultimately, the right amount of money has to be available to pay for resources of all types. This chapter deals with the techniques of ensuring the correct amount and timing of these three resources — material, equipment, and labor.

Close Management of Resources

Generally speaking, labor and equipment must be managed as tightly as possible to ensure their efficient use. A key element is the adequate and timely delivery of labor and material, important for two reasons: completion within contract limits, and a profit.

A project manager can greatly promote labor efficiency on a project by observing several principles. First, the rate and flow of work should be as orderly and even as possible. The project manager should try to schedule crews so that hiring is smooth and even in the beginning of the job, and that the overall size of the work force remains as constant as possible. Extreme peaks of demand should be avoided. Crews should be kept together, working on similar work for longer periods of time. It is very disrupting for skilled labor to have to jump around between different tasks, and to be alternately laid off and rehired. Productivity invariably suffers when these conditions occur. The ideal pattern for a given trade should look something like Figure 11.1, which shows an even build-up of labor, then a constant period for the middle of the project where most of the work gets done, followed by a smooth reduction of labor at the end of the task.

Adequate and timely resources (labor and material) must be supplied to keep the work flowing and to complete the project within the allotted contract time. The project manager can consult the estimate to find out how much total labor of a given type will be needed. However, the estimate seldom addresses the issue of *when* the labor will be needed, or when the peak period of need will occur. It also provides no information regarding conflicting demands for various resources as the job proceeds. This kind of information can only be derived by consulting the schedule.

It is very difficult to achieve an ideal labor use pattern (as shown in Figure

11.1), and peaks usually occur. However, using techniques based on critical path scheduling can significantly improve the chances for a successful project.

The Resource Management Process

Resource management is the decision-making process in which activities are prioritized and scheduled so that the expenditure of labor and/or equipment occurs in a desirable way. The best time to begin this process is early in the project. If resources are handled properly from the start, the project's chances of falling behind are minimized and there is no need to play "catch-up."

The logical starting point for resource management is the critical path schedule as calculated for the project. In the beginning, this schedule is based on physical logic. It typically has an uneven pattern of resource expenditure, but serves as the starting point for the resource management analysis and refinement.

To perform the analysis, the project manager must be able to project the pattern of expenditure for a particular resource. There are two basic views of this pattern:

1. The *Resource Profile* (sometimes called a *resource histogram*) — a compilation of the demand by time period for a given resource for a particular schedule. (See Figure 11.2 for a typical Resource Profile.)
2. The *Resource Summary Curve* (sometimes called an "S" curve) — the sum of the resource expenditures over time. See Figure 11.2.

Development of the Resource Profile

The development of the resource profile is a four-step process, illustrated in Figures 11.3 and 11.4. The sample used for this illustration is the building of the foundation up through the foundation walls. The resource being plotted for analysis is *formwork carpenters,* used for the forming and stripping work.

Step 1 — Calculation of Required Resources
This initial step is shown in Figure 11.3, which is a listing of each activity in

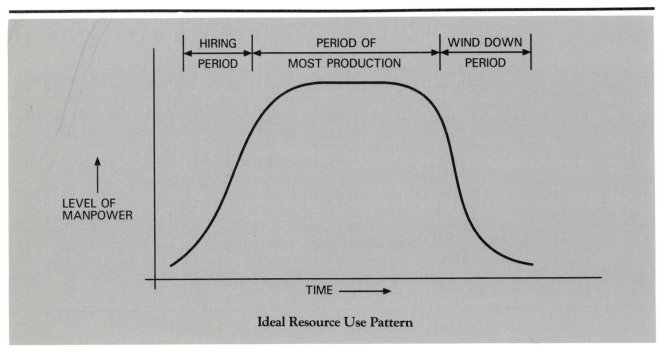

HIRING PERIOD   PERIOD OF MOST PRODUCTION   WIND DOWN PERIOD

LEVEL OF MANPOWER

TIME

Ideal Resource Use Pattern

Figure 11.1

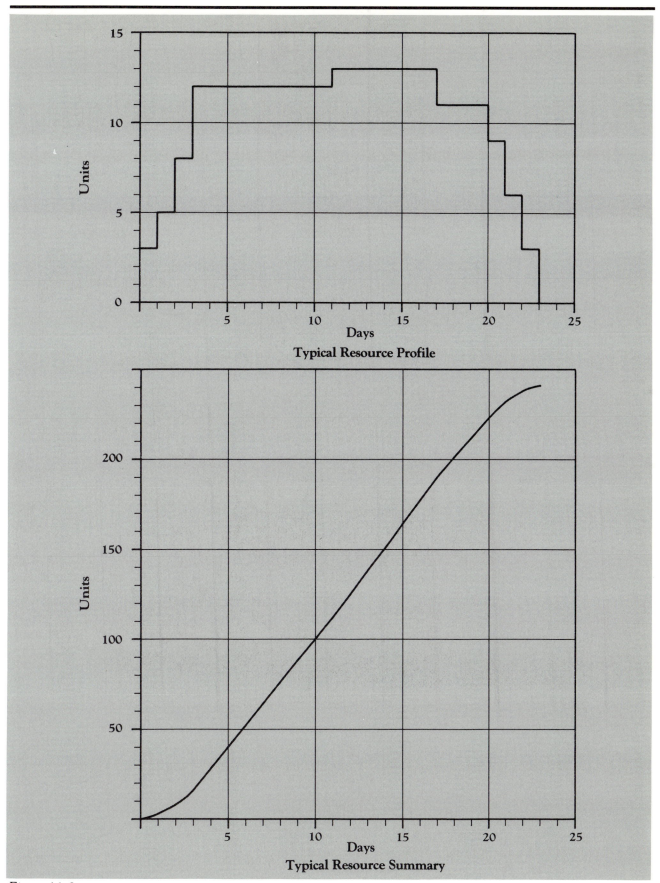

Typical Resource Profile

Typical Resource Summary

Figure 11.2

the sample network with quantities of work, man-hours per unit required, total man-hours, crew size, and number of days required. All of this information can be obtained from the estimate and/or a published source such as *Means Building Construction Cost Data.* The key numbers here are the total man-hours and crew size.

Step 2 — Distribution of Resources Across Activities

The next task is to assign the required resources to the appropriate activities. There are two methods:

1. Assigning the resources based on a rate of expenditure per day
2. Assigning resources based on a total amount of resources for the entire activity. The two methods are basically interchangeable.

For example, if the number of carpenter man-hours assigned per day is 4 men × 8 hours per day, the total number of carpenter man-hours for a six-day activity is:

6 days × 32 = 192 man-hours for the activity.

Alternately, the activity might have an assigned 192 man-hours, which can be translated back to four men per day.

192 man-hours/6 days = 32 man-hours per day, and
32 man-hours per day/8 hours per man = 4 men per day.

Both methods have advantages. If the time of the activity is likely to be varied, and the number of men adjusted up or down to vary the rate of progress, then it is probably better to assign total man-hours. If the number of men is to be fixed and the time of the activity is less definite, then assignment of number of men per day is preferable. In the case of the sample problem, the formwork carpenters are assigned by number of men per day since it is much simpler to calculate by hand. It can be seen in Figure 11.4 that the number of carpenters assigned per day to *form wall footings* is four, for *form column footings*, four, and so on.

Foundation Network Time Calculations							
	Quan.	Unit	MH/ Unit	Total MH Req'd.	Typ. Crew Size	Calc. Days Req'd.	Rounded Days Req'd.
Mass Excavation	9,100	CY	.008	73	1.5	6.1	7
Trench Excavation — Walls	116	CY	.107	12	3	.5	1
Trench Excavation — Columns	42	CY	.107	4	3	.2	1
Formwork: Wall Footings	1,962	SFCA	.050	98	4	3.1	4
Column Footings	1,578	SFCA	.058	92	4	2.9	3
Walls	11,172	SFCA	.092	1,028	12	10.7	11
Concrete: Wall Footings	31	CY	.640	20	8	.3	1
Column Footings	40	CY	1.280	51	8	.8	1
Walls	132	CY	.750	99	8	1.5	2
Reinf: Wall Footings	.8	TNS	15.250	12	2	.8	1
Column Footings	1.0	TNS	15.250	15	2	1.0	1
Walls	2.6	TNS	10.670	28	2	1.7	2
Stripping: Wall Footings	1,962	SFCA	.019	37	4	1.2	2
Column Footings	1,578	SFCA	.017	27	4	.8	1
Walls	11,172	SFCA	.030	335	4	10.5	11

Figure 11.3

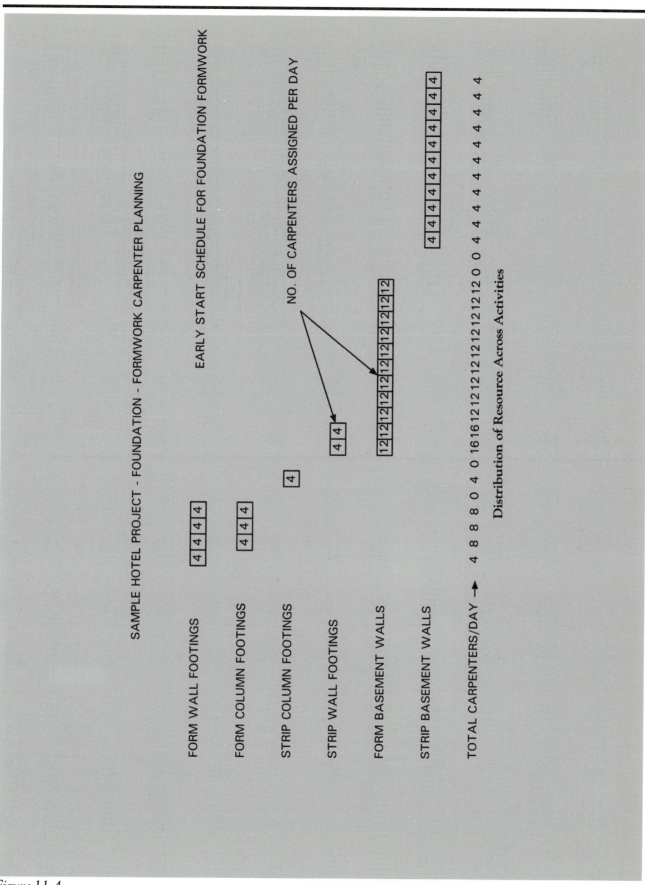

Figure 11.4

It should be noted that any unit of measure can be used for the resource. In this case, man-hours is used, but man-days are equally valid. In analyzing another resource, such as bulldozers, the unit of measure might be *equipment hours* or *equipment days*. It is best if the unit of measure remains consistent for all resources where possible in order to prevent confusion.

The schedule across which the resources are spread can be *early start, late start,* or a combination of both. The most commonly used schedule is *early start,* since it represents the desired schedule of most project managers.

Step 3 — Summarize Resource Expenditure by Time Period
This task is performed by totalling the figures (downward) at the end of each day of the job. In the sample problem (See Figure 11.4), it can be seen that the number of carpenters required per day for the first day is four, but for the second day the number required rises to eight, and stays at that level until the fifth day, at which point it drops to zero. This process is carried out across the project schedule until the end of the last carpenter activity, *strip basement walls.*

Step 4 — Plot the Resulting Profile
The tabular data arrived at in the previous step, while accurate, is not very useful to a project manager, who is better served by converting this information to a graphic representation. Figure 11.5, the resource profile, shows more clearly the periods of demand for formwork carpenters on the foundations.

The Value of Computerization

Resource management is a task well suited to computerization. Once the number of resources rises above one and the number of activities rises much above the level of the sample project, performing this task by hand becomes quite tedious and the possibilities for error increase. An example of computerized resource management is shown in Figures 11.6 and 11.7. The first is a bar chart showing the early start and late start times of the foundation activities previously plotted by hand. The second is a resource profile for *formwork carpenters,* using man-hours rather than man-days. The profile has the same shape in both figures and therefore shows the same pattern of expenditure.

Adjusting the Schedule to Improve the Resource Expenditures

The pattern of resource expenditure is important in that it has a very real effect on efficiency and productivity. For example, it is apparent that the gap in the profile at about March 7 would call for the laying off of some carpenters for several days or reassigning them to other parts of the job. Laying off carpenters for such a short period is not very good for morale, and is not a desirable alternative. Thus, the project manager must find another area in which they can work productively in the interim, until it is time to strip the basement wall forms.

Fortunately, once a diagram is created showing the pattern of expenditure for the resource, there is a way to determine how best to alter the schedule. The project manager has several options. His basic aim is to alter the schedule so that the pattern of use is as close as possible to the curve shown in Figure 11.1. To achieve this goal, he should try first to adjust non-critical activities within their float range to move peak periods of need into gaps in the diagram. The project manager can delay the start, extend the time, or work at varying rates to smooth out the flow. If the manipulation of non-critical activities does not work, then the project manager must begin altering the critical activities. A critical activity will probably have to be lengthened, and therefore the job as

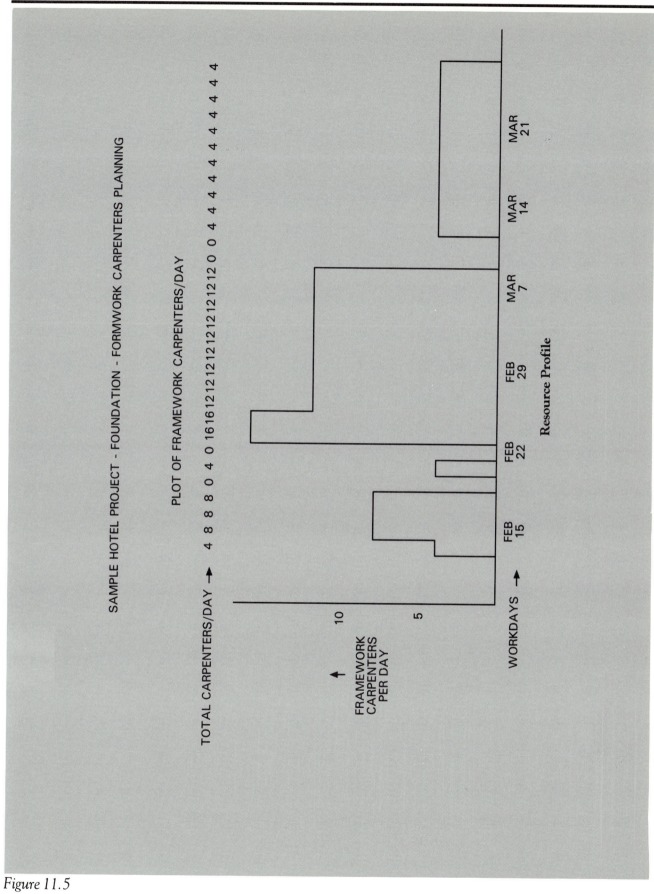

Figure 11.5

181

```
                                        01   08   15   22   29   07   14   21   28   04   11   18   25   02   09   16   23   30   06   13   20
                                        FEB  FEB  FEB  FEB  FEB  MAR  MAR  MAR  MAR  APR  APR  APR  APR  MAY  MAY  MAY  MAY  MAY  JUN  JUN  JU
ACTIVITY ID        TITLE                88   88   88   88   88   88   88   88   88   88   88   88   88   88   88   88   88   88   88   88   88

10010         MASS EXCAVATION     /EEEEE/
10020         EXCAV TRENCH/WALL *  . /
10040         REINFORCE WALL FO *  . E++L
10110         EXCAVATE COLUMN F *    E+++++++++++++++++++++L
10030         FORM WALL FOOTING *    /EE/
10140         REINFORCE COLUMN  *    E+++++++++++++++++++++++L
10120         FORM COLUMN FOOTI *    EEE++++++++++++++++++++++LLL
10050         PLACE WALL FOOTIN *     . E+++++++++++++++++++++++L
10130         PLACE COLUMN FOOT *       E+++++++++++++++++++++++L
10060         STRIP WALL FOOTIN *          //.
10150         STRIP FORMS COLUM *          EE+++++++++++++++LL
10080         REINFORCE BASEMEN *            EE++++++++LL
10070         FORM BASEMENT WAL *             /EEEEEEEE/
10090         PLACE BASEMENT WA *             //    /EEEEEEEE/.
10100         STRIP FORMS BASEM *                     /EEEEEEEE/.
10160         BACKFILL WALLS BA *                         . /EE/ /
19999         COMPLETE FOUNDATI *                         . /
```

Figure 11.6

182

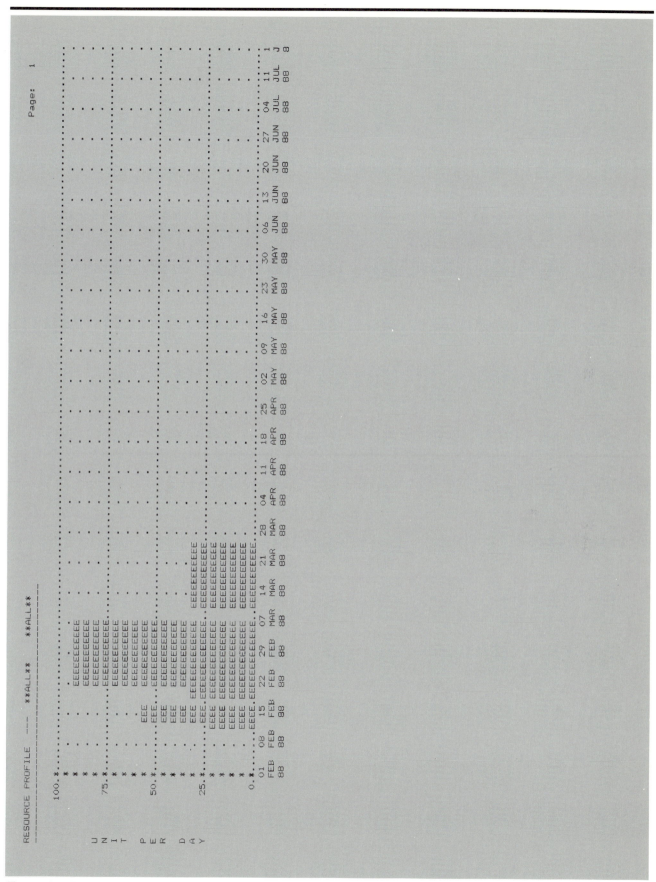

Figure 11.7

a whole is extended. This situation presents the project manager with one of his biggest challenges; trade-off decisions must be made between the value of smooth work flow and delay to the job.

A more complex example shows how schedule adjustment decisions might be made. Figure 11.8 is a bar chart showing the waffle slab structural work, beginning with *FRPS Columns — East, Bsmt to Main* and ending with *Complete Waffle Slab*. At first glance, this early start bar chart appears to have a reasonably smooth work flow. However, a resource profile for formwork carpenters (Figure 11.9) reveals some potential complications. Specifically, while there is a "base" need for about 54 man-hours per day, there are also five separate peaks in the need for carpenters, plus isolated periods of need after June 6. These anomalies must be resolved.

Comparing the bar chart to the resource profile, it can be seen that the periods of high demand occur when the various stripping activities are scheduled to take place. Further examination of the bar chart and resource profile reveals that the project manager is in luck; the activities that are causing the peak are non-critical!

Using the principle that it is preferable to reschedule non-critical activities rather than critical ones, the project manager might adjust the stripping activities as follows:

1. Extend the duration of the first three stripping activities from three days each to 15 days each. This adjustment is well within their allowable floats, so no delay is created.
2. Move the start of the next three stripping activities to a point after the last major forming activity is complete. This adjustment is within the float ranges.

The result of these adjustments can be seen in the new resource profile (Figure 11.10). The profile has been "flattened" considerably, by reducing the peaks from 75 man-hours per day to a reasonably smooth profile of less than 60 man-hours per day. Also the first "gap" has been closed and the total gap period has been reduced from eight to four days. The project manager should be able to work with this schedule without too much difficulty. Further analysis of the waffle slab combined with the tower structure that follows might close the gaps further, and result in an even better use of carpenters across the entire job.

Managing Cash Flow

In addition to managing the on-site resources of labor, material, and equipment, the contractor and project manager must also manage the flow of cash out of and into the job. All resources used to get the job built must ultimately be paid for in real dollars, and the project manager must ensure that sufficient funds are on hand to cover those obligations.

By nature, the payment system used in construction tends to create complications. In this process, the contractor submits an invoice for work done at the end of a regular period, typically every month. The owner pays at some later time. There is always significant delay in payment, usually not less than two to three weeks, and sometimes six weeks or more. Also, the owner usually deducts a retainage of five to ten percent of the value of the work done which is held until the end of the job. With this arrangement, the contractor is, to some degree, financing the owner. Meanwhile, the contractor must pay his own bills. Labor must be paid at the end of the week in which the work is done; material must be paid for very shortly after the end of the month in which the goods are received. In the case of labor, it is not uncommon for a

SAMPLE HOTEL SCHEDULE

REPORT DATE 7FEB88 RUN NO. 6

START DATE 4APR88   FIN DATE 23JUN88

DAILY BARCHART SCHEDULE

DATA DATE 1FEB88   PAGE NO.   1

DAILY-TIME PER.   1

............ACTIVITY DESCRIPTION............ ACTIVITY ID	OD	RD	PCT	CODES	FLOAT	SCHEDULE	04 APR 88	11 APR 88	18 APR 88	25 APR 88	02 MAY 88	09 MAY 88	16 MAY 88	23 MAY 88	30 MAY 88	06 JUN 88	13 JUN 88	20 JUN 88	27 JUN 88	04 JUL 88	11 JUL 88
FRPS W/S COLS,EAST, BSMT TO MAIN 20010	3				0	ERLY/LTE	EE														
FRPS W/S COLS-WEST, BSMT TO MAIN 20020	3				13	ERLY/LTE	.	EEE+++++++++LLL													
FRP WFFL SLBS-NE, MAIN LVL 20030	8				0	ERLY/LTE	. /EEEEEEE.														
FRPS COLS-NE, MAIN TO 2FLR 25010	3				17	ERLY/LTE		EEE++++++++++++++++LLL.													
FRP WFFLE SLBS-SE, MAIN LEVEL 20040	8				0	ERLY/LTE		/EEEEEEE													
FRPS COLS-SE, MAIN TO 2FLR 25020	3				17	ERLY/LTE			EEE++++++++++++++++LLL											
FRP WFFLE SLBS-SW, MAIN LVL 20050	6				0	ERLY/LTE			/EEEEE											
FRP WFFLE SLBS-NW, MAIN LVL 20060	6				0	ERLY/LTE				/EEEEE.										
FRP WFFLE SLBS-NE, 2FLR 25030	8				0	ERLY/LTE					/EEEEEEE									
STRP WFFLE SLBS-SW, MAIN 20090	2				22	ERLY/LTE					EE+++++++++++++++++++++++LL									
FRP WFFLE SLB-SE, 2FLR 25040	8				0	ERLY/LTE						/EEEEEEEE								
STRP WFFLE SLBS-NW, MAIN 20100	2				16	ERLY/LTE						EE+++++++++++++++LL								
STRP WFFLE SLB-NE, MAIN LVL 20070	2				8	ERLY/LTE							EE+++++++LL							
STRP WFFLE SLB-SE, MAIN LVL 20080	2				0	ERLY/LTE							./E							
STRP WFFLE SLB-SE, MAIN LVL 25050	2				0	ERLY/LTE							./E							
STRP WFFLE SLB-SE, 2FLR 25060	2				0	ERLY/LTE							./E							
COMPLETE WAFFLE SLAB 25999	0				0	ERLY/LTE							/							

Figure 11.8

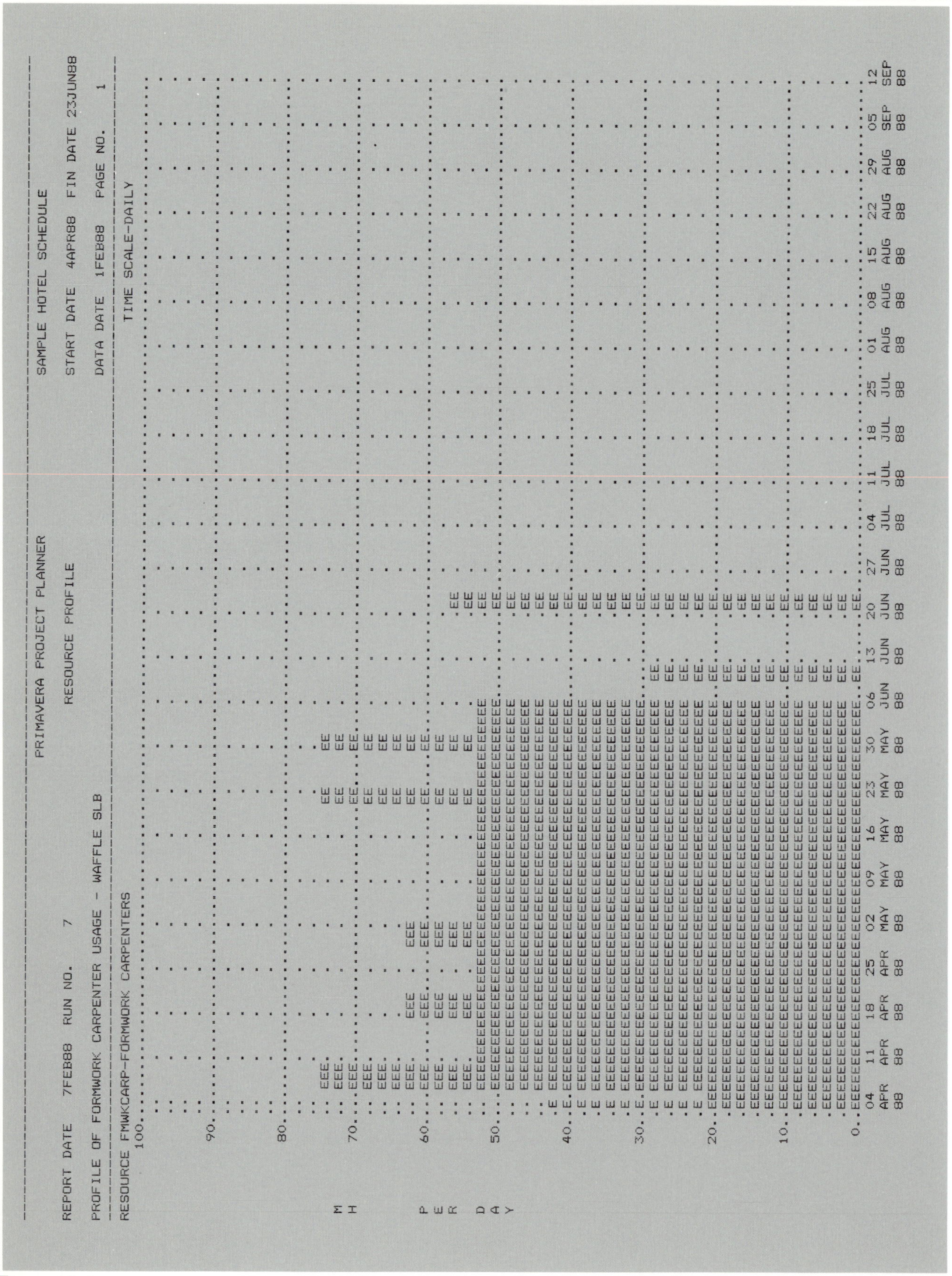

REPORT DATE 7FEB88 RUN NO. 7   RESOURCE PROFILE   START DATE 4APR88 FIN DATE 23JUN88

PROFILE OF FORMWORK CARPENTER USAGE - WAFFLE SLB   DATA DATE 1FEB88   PAGE NO.   1

RESOURCE FMWKCARP-FORMWORK CARPENTERS   TIME SCALE-DAILY

Figure 11.9

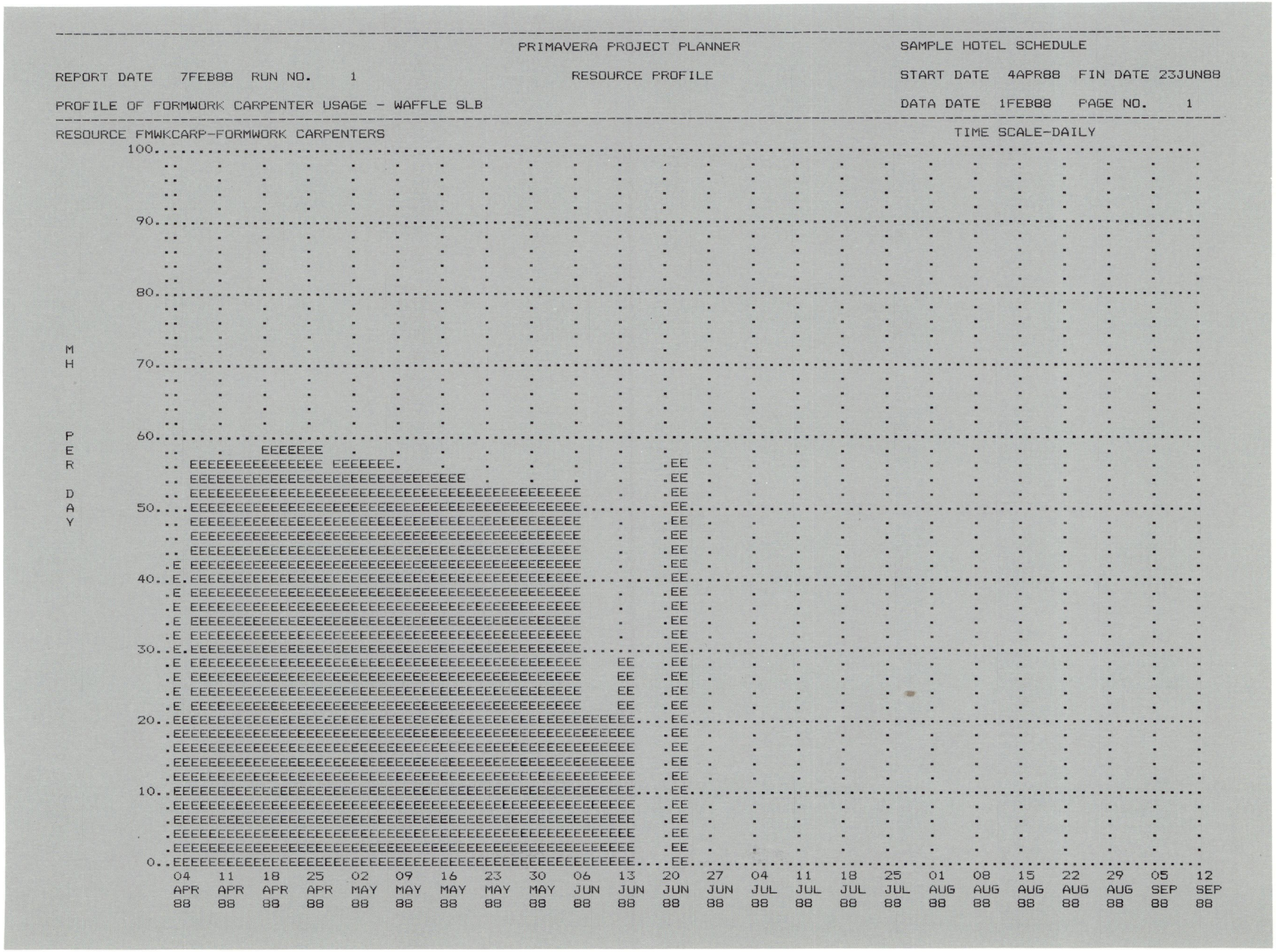

PRIMAVERA PROJECT PLANNER

SAMPLE HOTEL SCHEDULE

REPORT DATE 7FEB88 RUN NO. 1   RESOURCE PROFILE   START DATE 4APR88 FIN DATE 23JUN88

PROFILE OF FORMWORK CARPENTER USAGE - WAFFLE SLB   DATA DATE 1FEB88   PAGE NO. 1

RESOURCE FMWKCARP-FORMWORK CARPENTERS   TIME SCALE-DAILY

Figure 11.10

five-week lag to occur between the time a contractor has to pay workers and the time the owner pays the contractor for the work they performed.

The contractor must have cash on hand to cover his expenses. Access to that cash — in the form of borrowed money — has a cost in interest paid or internal cost of capital. The task for the contractor and project manager is to keep the cost of that borrowed capital to a minimum. This means having no more cash tied up in the project than is absolutely necessary. Any cash tied up in financing one project is unavailable for other projects, or for earning money in other areas such as interest-bearing investments.

A delicate financial balance makes it vitally important that the contractor carefully forecast and manage cash flow. Failure to do so can result in lower profits in the least extreme case, to failure of the company because the contractor takes on more work than there is cash to finance.

Forecasting and Managing Cash Flow

The techniques of managing *cash flow* are an extension of the *resource management* techniques just covered in this chapter. They involve using the critical path schedule to determine the timing of cash inflows and outflows on the job. The specific steps involved are as follows.

1. Assign a Cash Value to Each Activity

This is done by determining how much the project manager will spend getting each segment of work put into place. The cost can be broken down into labor, material, and equipment, but this is usually not necessary. An aggregate cost for the entire activity will suffice. Also, it is not necessary to determine at what point during the activity the money will have to be paid. Clearly, each of the three types of resources may be paid on a different basis. In practice, however, it is enough to simply spread the expenditure evenly across each activity.

The easiest way to find the cost of each activity is to take the units of work calculated for finding the duration of each activity, then multiply those units by the unit price for that work (as bid by the contractor). The sum of all the costs assigned to the activities must, of course, equal the total cost in the estimate and cost budget.

2. Plot the Resource Profile and Resource Summary Curves

Once the costs are assigned to the activities, the *Resource Profile* is plotted using the same technique previously covered for man-days or man-hours. This resource profile is then used to plot the *resource summary curve*, shown as the "S" shaped curve in Figure 11.11. The resource summary curve is simply the sum of the accumulated period costs as the project proceeds, and represents the forecast of when the money will be spent, or the *cash outflow curve*.

In practical terms, totalling the costs by hand is probably not a realistic undertaking if the project is more than a very small job. However, most computer CPM systems now have features that permit assignment of costs and summations for resource summary curves. These systems typically produce the curve in two forms: a tabular listing of the monthly cost and summary costs; and a plotting of the curve using a line printer. The tabular data is quite useful, the graph is often not. Most of the graphs allotted on this type of printer are too imprecise to be effectively read. It may, therefore, be best to transfer the data by hand to a piece of graph paper, or to a commercially available spreadsheet program, such as *Lotus 1–2–3* (Lotus Development Corp.) or *SuperCalc* 4 (Computer Associates), using the graphics functions of that type of program.

Plotting the cost curves of almost all construction projects results in curves of a shape similar to the one in Figure 11.11. This is characteristic of typical

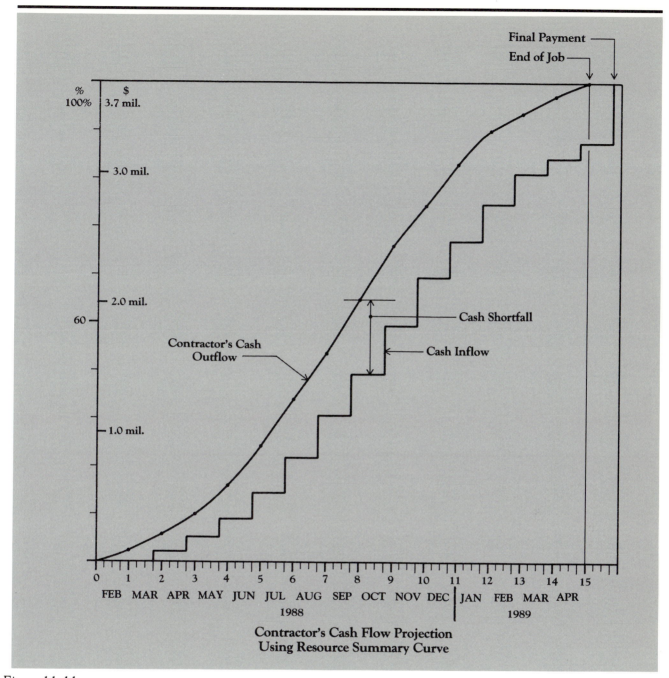

Figure 11.11

SAMPLE HOTEL PROJECT
CASH FLOW FORECAST

MO.	WEEK	PCT TIME	PCT WORK DONE	TOTAL WORK DONE	PAYMENT DUE	CASH SHORTFALL	
	1	1.667	.4	14800	0	(	14800)
FEB	2	3.333	1.2	44400	0	(	44400)
	3	5.000	2	74000	0	(	74000)
	4	6.667	2.4	88800	0	(	88800)
	5	8.333	3.2	118400	0	(	118400)
MAR	6	10.000	4	148000	0	(	148000)
	7	11.667	4.6	170200	79920	(	90280)
	8	13.333	5.8	214600	79920	(	134680)
	9	15.000	7	259000	79920	(	179080)
APR	10	16.667	7.6	281200	79920	(	201280)
	11	18.333	8.8	325600	193140	(	132460)
	12	20.000	10	370000	193140	(	176860)
	13	21.667	11	407000	193140	(	213860)
MAY	14	23.333	13	481000	193140	(	287860)
	15	25.000	15	555000	333000	(	222000)
	16	26.667	16	592000	333000	(	259000)
	17	28.333	18	666000	333000	(	333000)
JUN	18	30.000	20	740000	333000	(	407000)
	19	31.667	21.4	791800	532800	(	259000)
	20	33.333	24.2	895400	532800	(	362600)
	21	35.000	27	999000	532800	(	466200)
JUL	22	36.667	28.4	1050800	532800	(	518000)
	23	38.333	31.2	1154400	805860	(	348540)
	24	40.000	34	1258000	805860	(	452140)
	25	41.667	35.6	1317200	805860	(	511340)
AUG	26	43.333	38.8	1435600	805860	(	629740)
	27	45.000	42	1554000	1132200	(	421800)
	28	46.667	43.6	1613200	1132200	(	481000)
	29	48.333	46.8	1731600	1132200	(	599400)
SEP	30	50.000	50	1850000	1132200	(	717800)
	31	51.667	51.6	1909200	1451880	(	457320)
	32	53.333	54.8	2027600	1451880	(	575720)
	33	55.000	58	2146000	1451880	(	694120)
OCT	34	56.667	59.6	2205200	1451880	(	753320)
	35	58.333	62.8	2323600	1824840	(	498760)
	36	60.000	66	2442000	1824840	(	617160)
	37	61.667	67.4	2493800	1824840	(	668960)
NOV	38	63.333	70.2	2597400	1824840	(	772560)
	39	65.000	73	2701000	2197800	(	503200)
	40	66.667	74.4	2752800	2197800	(	555000)
	41	68.333	77.2	2856400	2197800	(	658600)
DEC	42	70.000	80	2960000	2197800	(	762200)
	43	71.667	81	2997000	2477520	(	519480)
	44	73.333	83	3071000	2477520	(	593480)
	45	75.000	85	3145000	2477520	(	667480)
JAN	46	76.667	86	3182000	2477520	(	704480)
	47	78.333	88	3256000	2763900	(	492100)
	48	80.000	90	3330000	2763900	(	566100)
	49	81.667	90.6	3352200	2763900	(	588300)
FEB	50	83.333	91.8	3396200	2763900	(	632700)
	51	85.000	93	3441000	2997000	(	444000)
	52	86.667	93.6	3463200	2997000	(	466200)
	53	88.333	94.8	3507600	2997000	(	510600)
MAR	54	90.000	96	3552000	2997000	(	555000)
	55	91.667	96.4	3566800	3116880	(	449920)
	56	93.333	97.2	3596400	3116880	(	479520)
	57	95.000	98	3626000	3116880	(	509120)
APR	58	96.667	98.4	3640800	3116880	(	523920)
	59	98.333	99.2	3670400	3236760	(	433640)
	60	100.000	100	3700000	3236760	(	463240)
	61	100.000		3700000	3236760	(	463240)
	62	100.000		3700000	3236760	(	463240)
	63	100.000		3700000	3700000		0

Figure 11.12

190

work patterns, and is so common that some companies use a "standard" "S" curve to predict cash flows, without plotting the specific pattern for each job. Surprisingly enough, this approach can be very effective, especially if the company builds relatively similar projects and develops a fairly consistent curve as they gain experience.

3. Plot the Cash Inflow Curve
The cash inflow is represented in Figure 11.11 as a series of steps upward as the payments come in from the owner. This illustration reflects certain assumptions that are fairly typical of actual practice. In this case, it is assumed that the owner will pay three weeks after the billing at the end of each month, and withhold 10% for retainage, to be paid with the final payment. The result, as shown in the diagram, is a payment schedule which is always behind the contractor's accumulated costs.

4. Determine the Cash Shortfall
The cash shortfall at any point is the difference between cash outflow and cash inflow. This represents the amount that the project manager must have available at a specific point to cover the bills that accrue as the project proceeds. Calculation of this value is another task that can easily be performed using a spreadsheet program. This value is shown as the last column in Fig. 11.12.

5. Update the Forecasts as the Job Proceeds
Finally, as various work activities are completed, the actual progress will vary somewhat from the original plan, and the curves will change accordingly. The information from the revised schedule should be transferred to the cash flow plots and acted on accordingly.

Practical Aspects of Resource Management

Some people in the construction industry claim that resource management techniques are relatively precise. This may be true on huge projects where the project managers have a large work force that can be moved around. However, on the vast majority of construction projects, precision is extremely difficult to attain. The techniques of resource management are at best an approximate, but still very valuable, tool. They can be used to smooth the rough edges and improve the overall project plan. When using resource management techniques, it is helpful to observe the following guidelines.

1. As noted, these techniques offer approximate solutions, so perfection should not even be attempted, much less expected. They are also very much trial and error; be prepared to take several passes at leveling and smoothing out a resource such as formwork carpenters.
2. Give priority to the critical and near critical activities, ensuring that they receive enough labor and equipment to stay on schedule. The non-critical activities can be adjusted as necessary, with the activities that have the most float having the greatest flexibility. Be careful, however, not to let the non-critical activities slip by unnoticed.
3. Constant monitoring of labor and equipment use is necessary to ensure compliance with the plan, or to notice if the plan is not working. Toward this end, be prepared to revise and re-plan frequently and whenever necessary.
4. It is probably not realistic to look ahead more than a few weeks at a time and attempt to work things out in detail. The advantage of this approach is that it is possible to apply what is learned in the early stages of a job to the later stages.
5. Computers are great for plotting the data, but are terrible for making decisions. Some computer programs have leveling algorithms which are

purported to provide the most efficient redistribution of labor or other resources. As a practical matter, these programs only work on the largest of jobs. Even then, their effectiveness is questionable. These leveling programs are also extremely complex. Thus, one can use them in good faith without completely understanding what they are calculating.

6. The best use of computers is plotting resource profiles. The project manager can take the computer printout sheet, mark it up with his ideas for change, put the changes back into the computer, and find out the result. The computer allows this "what-if" game, where a number of options can be tried out very easily, leading to a better overall solution by the project manager.

12

Time/Cost Trade-offs

12

Time/Cost Trade-offs

So far, the emphasis of this book has been on planning a job and then keeping it on schedule and on budget. However, the time will inevitably come when the duration of a job must be speeded up, or shortened, for some reason. This speeding up process is often referred to as, "crashing," "project expediting," or "accelerating." By whatever name, it is a difficult task for a construction manager. The challenge is making sure that the acceleration is carried out with the least amount of extra cost and disruption to the project as a whole.

Fortunately, there are specific, sound techniques for accomplishing the speeding up process. The critical path schedule itself can be a very useful tool for carrying out this task. It can serve as a basis for finding out why the project is behind, and even more importantly, for examining alternatives for "crashing" the job. These techniques, the subject of this chapter, are the project manager's best means of reducing project time with the least amount of panic, disruption, and cost increase.

Reasons for Reducing Overall Project Time

There are many factors that may encourage or force a reduction in project time. Most commonly, a schedule, as originally developed, is simply too long for the allowed project time, and must be shortened. Other motivations might be the advantages of completing a project early in order to earn bonus payments, or to avoid the complications of upcoming winter weather.

Much more common, however, is the situation in which the project manager finds the job behind schedule due to events that have occurred during construction. For example, subcontractors can be late in completing work, materials may not be delivered on time, or bad weather may hold up progress. As a result, the projected time of completion — after updating and entering progress — is beyond the original allowed date. This is a much more difficult case, since the original condition causing the delay is often still present on the job. Also, the project manager is often working under pressure from the owner, and the cost of accelerating the job must be measured against the cost of liquidated damages and loss of professional reputation.

In either case, the task of intelligently and calmly reducing project time is a difficult one. In fact, it is almost impossible to accomplish effectively without using techniques such as those described in the following pages.

Relationship of Project Cost to Time

Before delving into the specific techniques, it is useful to look at what happens to project cost when the time is reduced. The situation can be stated very simply: if project time is reduced, project cost almost always goes up. The problem for the project manager is to keep that cost increase to a minimum.

The graph shown in Figure 12.1 illustrates this phenomenon of reduced time and increased cost. Most people who are familiar with the construction process can readily identify with this problem. For example, if a project's duration is reduced by the use of personnel overtime, not only does the hourly cost of labor increase, but the productivity of that labor often decreases. The result is a "double whammy," both parts of which increase the overall cost of construction. An alternative to overtime — using more crews in the same space — has the disadvantage of causing interference problems. Any experienced construction supervisor is aware of the difficulties associated with supervising more than a usual number of workers and crews.

How to Reduce Project Time

The project manager who must reduce project time has these basic concerns:
1. Where and how should the project be reduced?
2. How much will each increment of time reduction add to the overall project cost?

The answer to the first question is relatively easy; the answer to the second is more difficult.

Look at the Critical Path

When it comes to deciding where and how to reduce the project time, the answer is: look at the critical path. Since the definition of the *critical path* is that group of activities which must be completed on time in order to complete the project on time, then it follows that any shortening of overall time must occur in that same path. It also follows that reducing time in any other part of the network (i.e., the non-critical activities) will only result in gaining more float time.

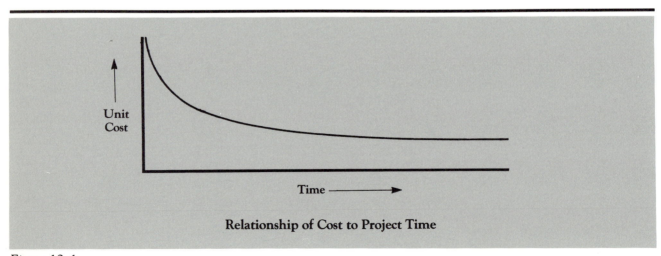

Relationship of Cost to Project Time

Figure 12.1

The easiest way to find the critical path is to select all activities with a *Total Float* equal to zero, and sort them by *Early Start*. An example of this type of run is shown in Figure 12.2. Also useful is a Total Float — Early Start (TF-ES) sorted schedule of the project, although this type tends to be very long since it displays all the non-critical activities as well. It can be seen from this run that the critical path for the hotel project runs through the foundation, then the waffle slab structure, and finally through the interior sequence for the top floor of the tower. A more detailed examination and analysis of the critical path can be performed using smaller segments of the work e.g., the waffle slab or tower sequence alone.

Once the critical path has been determined, the shortening process can be approached in one of two ways, or using a combination of both. These are:

1. Changing the network logic, and
2. Shortening the individual activities

Changing the Network Logic   The simplest and least costly way of shortening the project time is to simply introduce lags into the logic. As noted in earlier chapters, it is best to use a conservative approach when developing logic initially. This means building in primarily finish-to-start relationships. In reality, most construction sequencing is not so conservative. In any case, some overlap is usually possible between many of the activities on a job. If the original schedule is developed conservatively, then the project manager can look back through the critical path, and selectively decide to overlap certain activities in order to regain lost time. An example is shown in Figures 12.3 and 12.4 in which a segment of the logic from the sixth floor interior of the sample hotel is shown first with conservative logic, and then with lags introduced. In this example, the interior activities are originally scheduled in such a way that one trade must be clear before the next is introduced to the floor. In fact, the building is long enough that one trade can be halfway down the floor by the time the next trade is started. In this case, it is possible to save seven days, with little, if any, interference between the various trades.

Shortening Individual Activities   The second method for shortening the critical path consists of finding ways to reduce the time for individual activities. This usually means applying more labor and equipment to the work. Many instances that require shortening lend themselves quite well to this solution. For example, in the interior of the main floor of the sample hotel, hanging gypsum wallboard occurs over a large floor area, and extra crews can easily be added at little or no extra cost. Twelve carpenters could operate just as efficiently as eight, and the time for hanging gypsum wallboard would be cut by one third. It should be noted again, however, that increasing labor by using overtime is usually not a cost-effective solution, and this measure should be used with discretion. Not only does overtime increase the labor cost per hour, but it has also been found that the output per man-hour decreases after the eighth hour of work. The result is an increase in unit cost installed.

Not all cases are so simply calculated. For instance, in situations where the rate of production is dependent upon a balanced crew, the application of more labor will not increase production. An example of such an activity is forming slabs over the garage and main floor areas. Similarly, work that is dependent upon machinery cannot be speeded up by the application of extra labor. An example is the placement of steel, which requires the use of cranes. Only the application of extra cranes can speed up progress, and this is sometimes not possible due to space restrictions. Also, trying to add extra

ACTIVITY ID	RD	PCT	TITLE	EARLY START	EARLY FINISH	TF
10010	7	0	MASS EXCAVATION	1FEB88	9FEB88	0
10020	1	0	EXCAV TRENCH/WALL FOOTINGS	10FEB88	10FEB88	0
10030	4	0	FORM WALL FOOTINGS	11FEB88	16FEB88	0
10050	1	0	PLACE WALL FOOTINGS	17FEB88	17FEB88	0
10060	2	0	STRIP WALL FOOTINGS	18FEB88	19FEB88	0
10070	11	0	FORM BASEMENT WALLS	22FEB88	7MAR88	0
10090	2	0	PLACE BASEMENT WALLS	8MAR88	9MAR88	0
10100	11	0	STRIP FORMS BASEMENT WALLS	10MAR88	24MAR88	0
10160	4	0	BACKFILL WALLS BASEMENT	25MAR88	30MAR88	0
19999	0	0	COMPLETE FOUNDATION	31MAR88	31MAR88	0
20010	3	0	FRPS W/S COLS,EAST, BSMT TO MAIN	31MAR88	4APR88	0
20030	8	0	FRP WFFL SLBS-NE, MAIN LVL	5APR88	14APR88	0
20040	8	0	FRP WFFLE SLBS-SE, MAIN LEVEL	15APR88	26APR88	0
20050	6	0	FRP WFFLE SLBS-SW, MAIN LVL	27APR88	4MAY88	0
20060	6	0	FRP WFFLE SLBS-NW, MAIN LVL	5MAY88	12MAY88	0
25030	8	0	FRP WFFLE SLBS-NE, 2FLR	13MAY88	24MAY88	0
25040	8	0	FRP WFFLE SLB-SE, 2FLR	25MAY88	3JUN88	0
20080	2	0	STRP WFFLE SLB-SE, MAIN LVL	20JUN88	21JUN88	0
25050	2	0	STRP WFFLE SLB-SE, MAIN LVL	20JUN88	21JUN88	0
25060	2	0	STRP WFFLE SLB-SE, 2FLR	20JUN88	21JUN88	0
25999	0	0	COMPLETE WAFFLE SLAB	22JUN88	22JUN88	0
33010	6	0	FRPS SHR WALLS-N	22JUN88	29JUN88	0
33030	7	0	FRP SLAB-N	30JUN88	8JUL88	0
33040	6	0	FRPS SHR WALLS-S	11JUL88	18JUL88	0
33060	7	0	FRP SLAB-S	19JUL88	27JUL88	0
34010	6	0	FRPS SHR WALLS-N	28JUL88	4AUG88	0
34030	7	0	FRP SLAB-N	5AUG88	15AUG88	0
34040	6	0	FRPS SHR WALLS-S	16AUG88	23AUG88	0
34060	7	0	FRP SLAB-S	24AUG88	1SEP88	0
35010	6	0	FRPS SHR WALLS-N	2SEP88	9SEP88	0
35030	7	0	FRP SLAB-N	12SEP88	20SEP88	0
35040	6	0	FRPS SHR WALLS-S	21SEP88	28SEP88	0
35060	7	0	FRP SLAB-S	29SEP88	7OCT88	0
36010	6	0	FRPS SHR WALLS-N	10OCT88	17OCT88	0
36030	7	0	FRP SLAB-N	18OCT88	26OCT88	0
36040	6	0	FRPS SHR WALLS-S	27OCT88	3NOV88	0
36060	7	0	FRP SLAB-S	4NOV88	14NOV88	0
37010	6	0	FRPS SHR WALLS-N	15NOV88	22NOV88	0
37030	7	0	FRP SLAB-N	23NOV88	1DEC88	0
37040	6	0	FRPS SHR WALLS-S	2DEC88	9DEC88	0
37060	7	0	FRP SLAB-S	12DEC88	20DEC88	0
37080	3	0	STRP SLAB-S	4JAN89	6JAN89	0
37090	7	0	HANG CURTAIN WALL - 6 FLR	9JAN89	17JAN89	0
66020	8	0	INSTALL DOOR FRAMES & STUDS	18JAN89	27JAN89	0
66050	7	0	ROUGH IN ELECTRIC	30JAN89	7FEB89	0
66060	7	0	HANG & TAPE GWB	8FEB89	16FEB89	0
66070	2	0	ROUGH PAINT	17FEB89	20FEB89	0
66080	2	0	SPRAY ACOUS PLSTR	21FEB89	22FEB89	0
66090	3	0	INSTALL CER. TILE	23FEB89	27FEB89	0
66120	4	0	INSTALL COUNTER TOPS	28FEB89	3MAR89	0
66140	6	0	INSTALL VWC	6MAR89	13MAR89	0
66150	5	0	FINISH PLUMBING	14MAR89	20MAR89	0
66170	5	0	INSTALL BATHROOM ACCESSORIES	21MAR89	27MAR89	0
66180	2	0	INSTALL SPRNKLR HEADS	28MAR89	29MAR89	0
66200	4	0	LAY CARPET	30MAR89	4APR89	0
66210	3	0	HANG DOORS & HARDWARE	5APR89	7APR89	0

Figure 12.2

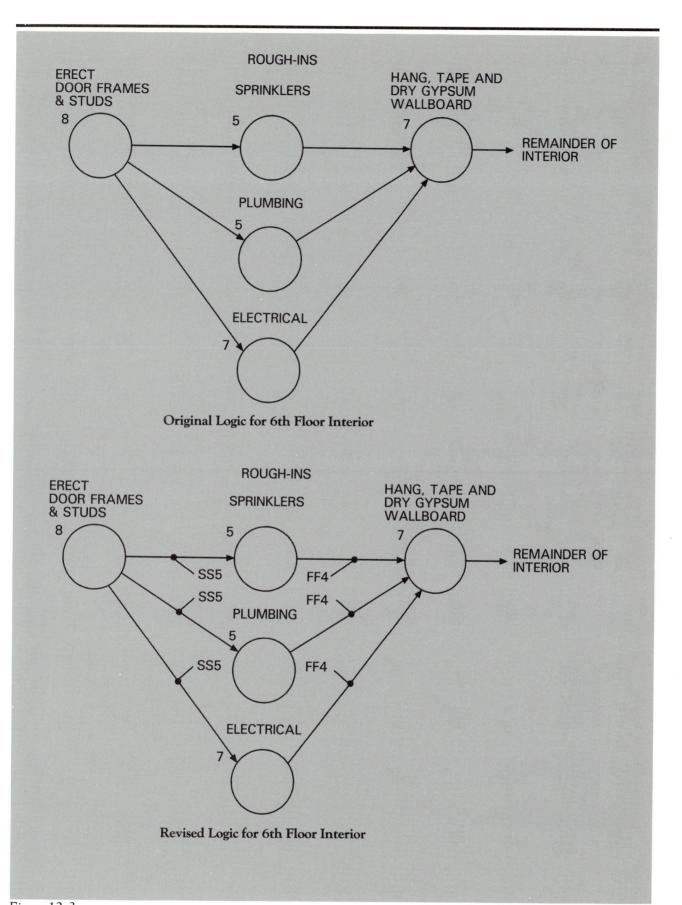

ROUGH-INS

ERECT DOOR FRAMES & STUDS

SPRINKLERS

HANG, TAPE AND DRY GYPSUM WALLBOARD

8

5

7

REMAINDER OF INTERIOR

PLUMBING

5

ELECTRICAL

7

Original Logic for 6th Floor Interior

ROUGH-INS

ERECT DOOR FRAMES & STUDS

SPRINKLERS

HANG, TAPE AND DRY GYPSUM WALLBOARD

8

5

7

REMAINDER OF INTERIOR

SS5

SS5

FF4

FF4

PLUMBING

5

SS5

FF4

ELECTRICAL

7

Revised Logic for 6th Floor Interior

Figure 12.3

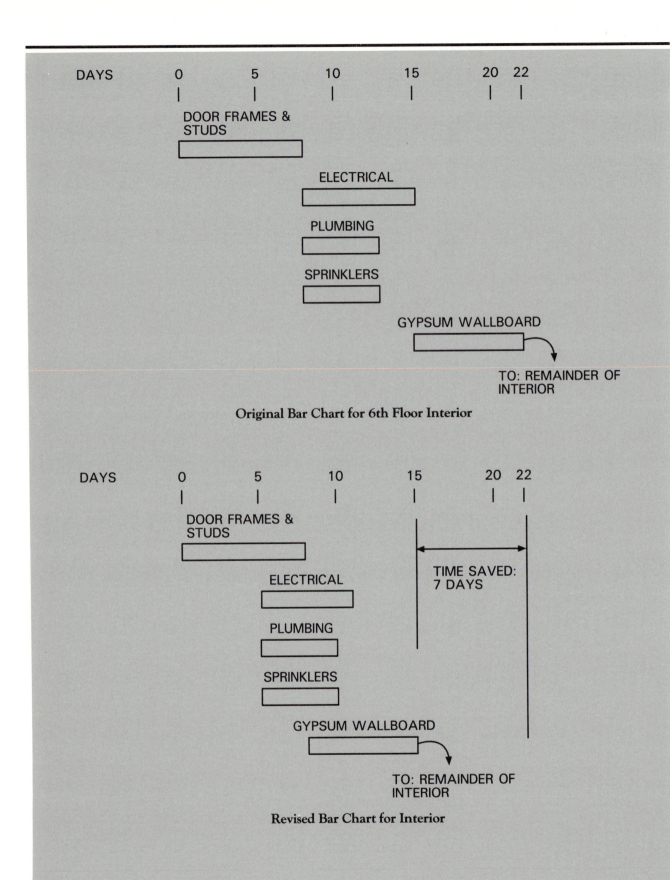

Original Bar Chart for 6th Floor Interior

Revised Bar Chart for Interior

Figure 12.4

machinery can sometimes result in interference which can slow down all crews, thus causing unacceptable increases in cost.

It is of course possible to change logic and add labor or equipment at the same time. Resources can be added in the form of more people on the crew, or additional crews. Logic changes can be simple or extensive. Figure 12.5 is an example of a more complex case — a significant shortening of the tower structural sequence. (Column work is not shown in order to simplify the example.) The original logic called for one crew, identified as "Crew A" on the diagram. The sequence of their work would be shear walls to slab on one side of the floor, then shear walls and slab on the other side, after which they would proceed to the next floor. This sequence is simple and totally sequential for the crew, and using this plan, a floor is completed every 26 working days.

The other sequence shown calls for a second crew of equal size (identified as "Crew B"), which begins after Crew A has completed its first set of shear walls. Crew A, instead of doing slab work, installs the shear wall on the other end of the floor, while Crew B starts on the slab that will rest on the shear walls just completed by Crew A. Both crews then proceed up the building, with Crew A preceding Crew B on each segment.

The total manpower is doubled in this case. Five days are saved on the first forming cycle, and 17 days saved on the second cycle. Each subsequent cycle will be completed even earlier because not only is each shortened from 26 to 21 days, but there is now an overlap in floor work of seven days.

Overall, this revised sequence would significantly shorten the project. It would also increase the cost to some degree, as well as complicating management tasks somewhat for the superintendent and project manager. For example, additional formwork would probably have to be purchased or rented because of the faster sequence. The column sequence might have to be faster as well. Also, because so many additional people are on the crews, coordination and timing must be more closely monitored. Since the concrete curing time becomes more critical, the structural engineer should be consulted to ensure that the sequence is safe. Overall, the project will move considerably faster, and the additional expenditure will not jeopardize the overall cost limits.

Practical Considerations

When it comes to the practical business of reducing project time, the following guidelines should be observed.

First, it pays to keep the process as simple as possible. A project manager seeking to shorten a critical path should choose certain key elements that can be shortened and then concentrate the reduction efforts on those elements in the most careful manner possible. The elements that are chosen for shortening should be those most amenable to cheap time reduction, i.e., those that are easiest to overlap, or to which more labor can be applied with the least problems. The activities involved in the shortening should then be monitored more closely. A guaranteed way to drastically increase project cost is to try to "crash" across the board. What typically happens is that coordination becomes nearly impossible, and the entire job ends up essentially out of control.

Second, one should recognize that the additional cost of reducing project time cannot be precisely predicted. The graph shown in Figure 12.1 is only representative of the general case. Very little data exists to provide a project manager with precise data on how much the unit cost of an item will increase if activity duration is decreased by 10 or 20%. The best the project manager

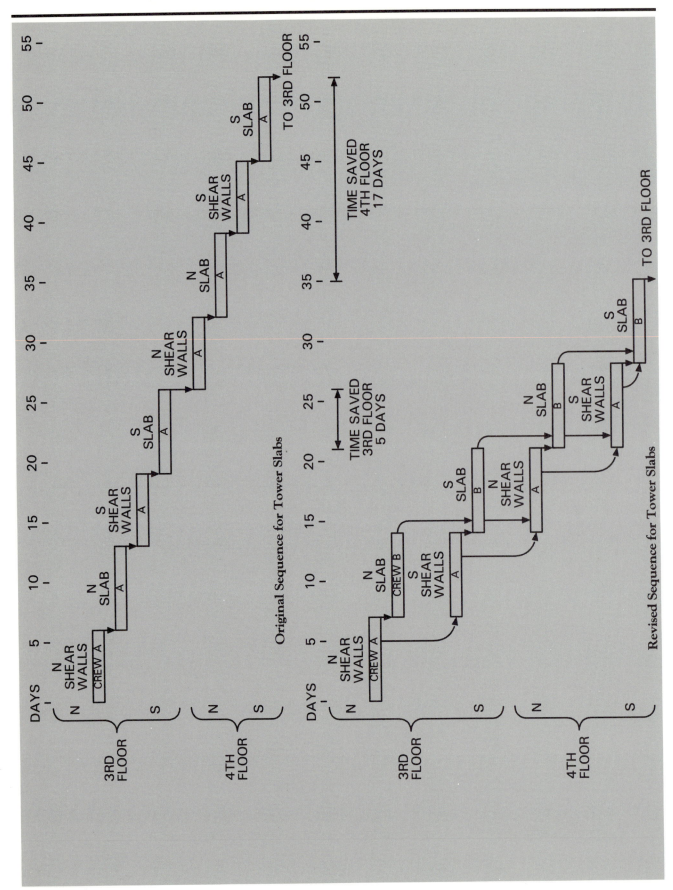

Original Sequence for Tower Slabs

Revised Sequence for Tower Slabs

Figure 12.5

202

can do is to use educated judgment and keep the number of affected activities to a minimum.

Third, it is important to recognize the fact that time compression is often done under severe pressure. A job that is drastically behind schedule does not tend to result in a pleasant working atmosphere, and a project manager in this situation is under enormous pressure from owner and company management alike to get things back on track. This situation limits the ability of the project team to implement complex solutions effectively, and the best approach is often the simplest.

13

Project Cost Control

13

Project Cost Control

So far we have concentrated primarily on the methods for keeping a project on schedule. The other half of project management, keeping the project within budget, is the subject of this chapter. Cost control tends to be the priority of many companies. For example, subcontractors who do only one type of work have virtually no need to run scheduling systems; their schedules are set by the general contractor. Cost control, on the other hand, is still of vital importance.

Cost control is typically an easier task for the project manager than time control, but it is no less important. A good part of the project cost control function is performed within the overall accounting system of a construction company, and is not a burden to the project manager. This chapter covers the role of the project manager in the cost control process, and what that individual needs to do and know to effectively control costs.

Construction cost control involves performing the same functions as any business selling a product for a profit. This production cost control is just as critical for construction as it is for other businesses. Cost control for construction has the following additional complications. First, the elements that influence cost variations in construction can be highly volatile. Most manufacturing businesses are not constantly producing a different product; the size of the product relative to the total company size is not so huge as it is in construction; and weather is not the significant factor that it is in construction. These elements are always present in the construction business, and they make an effective, functioning cost control system an absolute necessity.

In construction, cost factors can be broken down into four categories: labor, material, equipment and subcontractors. Of these four, material and subcontractor costs are relatively easy to control. Material costs tend to be fairly predictable, and subcontractor costs are defined at the time of the bid and job buy-out. Labor and equipment are, however, an entirely different matter. These two factors constitute the greatest risk for large cost overruns and, in many cases, have the potential for bankrupting the project and even the company.

The real key to controlling labor and equipment costs lies in the *feedback cycle* concept described in Chapter 2. To keep costs under control, the project manager must set targets for the various categories, closely monitor the performance of the crews doing the work, and when a deviation in cost performance is found, he must act positively and quickly to bring about corrective action. This monitoring must be done at the job site. The principle of being close to the work is a basic tenet of management, and nowhere more valid than in construction. The emphasis must be on aggressive job site tracking and follow-up, not monitoring from the home office. The Project Control Cycle described in Chapter 2 provides the basis for operating a cost system, just as it does for scheduling.

Project Cost Coding Systems

All cost control systems depend on a *project cost code.* This system of classifying costs and types of work is essential if the numbers collected in the job tracking process are to be useful to the project manager. The key to keeping informed of job progress is the comparison of actual performance to the targeted progress. The project manager should be able to look at a specific category of work separate from all others, and make meaningful judgments about cost and performance. This is not possible if he cannot clearly isolate finish work from concrete work.

Information on cost and performance is collected and represented in the form of lots of numbers — man-hours worked, dollars spent, cubic yards of concrete placed, etc. Consequently, there should be a means of classifying the numbers into categories meaningful to the project manager. A coding system provides such a means and ensures that "apples are compared to apples." When a man-hour is expended on concrete work, that man-hour will be recorded as an expense of concrete work and will not be assigned to some other category of work. When an accounting of expenditures has been made, the totals and other information can then accurately be compared to the target figure.

Elements of a Project Coding System

Every project coding system must serve a number of functions. Over the years, systems have evolved to meet these various needs. While there is no such thing as one standard project cost coding system, most numbering schemes contain all or most of the following elements.

Project Number All systems identify each project as being a separate entity. Usually the project number corresponds to a specific contract that the company has undertaken. The project number also frequently includes a number identifying the year in which the project was started. These two elements are derived from tax accounting requirements (based on the requirements of the I.R.S.) i.e., the need to identify income and expenditures by project and fiscal year. A typical project number might be 8823. This designation might indicate the company's 23rd job, started in fiscal year 1988. Some companies simply number projects sequentially from the first job obtained, keeping a list of when each job started. In this case, job number 435 would be the 435th job since the company started.

Area, Job Type or Other Subclassification Code. If the company's geographic spread or variation in types of work is significant, a code might be added to clarify these distinctions. For example, say a company does union work in the Midwest and in California, but (through a subsidiary) does work in the Sunbelt that is non-union. Clearly, each division or subsidiary is affected by different cost factors. Since it would not be advisable to compare costs from one area to the other, each area would have a separate code. Another example is a company with different divisions performing very different types of work in the same geographic area. Power plant work, for example, is significantly different from building construction and should be recorded separately. If such variations do not exist, then this part of the overall project cost code may not be necessary. A contractor working only in the Los Angeles Basin would probably not need to distinguish between overall conditions of work in El Segundo and Orange County.

Work Type This is probably the most essential part of the code from the standpoint of detailed project cost control. It is this number that separates different materials and trades on the job. it also furnishes the basis for identifying crews and ongoing work tasks, and for determining whether or not these controllable areas need attention. The starting point for a work type code is almost always the classifications of the MASTERFORMAT system. The "Broadscope" headings outlining the MASTERFORMAT system are reproduced in Appendic C. The code is hierarchial, e.g., under *concrete* there are sub-categories such as *concreting procedures, concrete formwork,* etc. These subdivisions are broken down further into more detailed classifications. *Concrete formwork,* for example, further subdivides into *wood forms* and *formwork accessories.* One point to note, however, is that the MASTERFORMAT classifications may not precisely cover all possible situations. A case in point is if the contractor wishes to make a distinction between *erecting and stripping forms,* additional categories must be developed.

Type of Expense Category Finally, each expenditure should be recognized as one of the four types of spending: labor, material, equipment, and subcontractor costs. These classifications are sometimes called *cost distributions.*

Detail in a Cost Code System
How much detail should be built into the system? On the one hand, more detail provides more information about precisely where cost overruns are occurring. On the other hand, a greater amount of detail makes the system bureaucratic and heavy with paperwork. If the system is too detailed, the likelihood of error actually increases. A system that demands excessive detail may be resented by the field personnel who have to fill in the forms that provide the system with its input. The answer to the problem is that it is better to have a system that works well and accurately, even at the expense of slightly less precise information. A tremendously detailed system may sound great from the standpoint of telling everything there is to know about the project, but it is no good if it is not accurate. Figures 13.1a and 13.1b show the kinds of extremes which are possible on coding system detail.

Other Points about Cost Codes
Several practical points should be noted by the company or project manager considering a new cost system, or changing an old one. First, make sure the system fits the company's needs. This is often a major factor for a company buying a new computer and accounting system, as the code system may be changed at the same time. In fact, if the present code system works fine and the computer system is being changed for capacity reasons, then a strong case should be made for keeping the old classification system. A coding system is a lot more effective when company personnel are used to using it, and changing them over entails significant risk and cost. When upgrading, the direction to take is *evolutionary,* not revolutionary.

Second, the coding system should be well explained in the company's field operations manual. Each code should be thoroughly and clearly explained, and the manual should be readily available to everyone who has to use it.

Specific Tasks in Project Cost Control

The specific steps or tasks that the project manager and his team must undertake fit within the schematic of the Project Control Cycle described in Chapter 2. The sequence for cost control is shown in Figure 13.2. These steps are described in detail as follows.

Estimate the Job

The estimate is the basis for the project's cost goals. As such, it represents the limits of spending which must be met by the project manager if the project is to realize the profit anticipated at the time the contract was obtained. An example of a typical, detailed estimate for the foundation and waffle slab concrete work of the sample hotel project is shown in the Appendix. A summary of that estimate is shown in Figure 13.3.

Unfortunately, estimates are usually not arranged in a manner immediately suitable to cost control purposes. The problem is one of classification of work subcategories. In an estimate, the job is broken down in such a way as to simplify the complications of bid day. It is not designed as an aid to control work in the field. The estimate may be more or less detailed than the cost code system of the company, but in any case, must almost always be reworked to make the information useful for cost control.

Simple Cost Coding System		
031.00	Formwork	
	031.10	Formwork Material and Accessories
	031.20	Fabricating, Erecting, Stripping and Moving Formwork
	031.21	Foundations
	031.22	Slabs on Grade
	031.23	Columns
	031.24	Walls
	031.25	Elevated Slabs
032.00	Reinforcing	
	032.10	Reinforcing Materials and Accessories
	032.20	Sorting, and Placing Reinforcing
	032.21	Foundations
	032.22	Slabs on Grade
	032.23	Columns
	032.24	Walls
	032.25	Elevated Slabs
033.00	Placing and Finishing	
	033.10	Materials and Accessories
	033.20	Concrete Placement
	033.21	Foundations
	033.22	Slabs on Grade
	033.23	Columns
	033.24	Walls
	033.25	Elevated Slabs
	033.30	Concrete Finishes
	033.31	Horizontal
	033.32	Vertical
	033.33	Curing

Figure 13.1a

031.0000   Concrete Formwork
     031.1000   Fabricating and Erecting
          031.1100   Foundations
               031.1110   Spread Footings
               031.1120   Pile Caps
               031.1130   Equipment Pads
          031.1200   Slabs on Grade
               031.1210   Sidewalks
               031.1220   Stairs on Grade
               031.1230   Exterior Curb and Gutter
               031.1240   Concrete Paving
               031.1250   Special Finishes
          031.1300   Columns
               031.1310   Conventionally Formed
               031.1320   Prefabricated Forms
               031.1330   Disposable Forms
          031.1400   Walls
               031.1410   Conventional Forming
                     031.1411   Straight Walls
                     031.1412   Curved Walls
               031.1420   Gang Formed
                     031.1421   Straight Walls
                     031.1422   Curved Walls
               031.1430   Special Surface Finish Formwork
          031.1500   Elevated Slabs
               031.1510   Flat Slabs
               031.1520   Drop Panel Slabs
               031.1530   Pan or Dome Slabs
               031.1540   Flying Forms
               031.1550   Metal Deck Forming Systems
     031.2000   Stripping, Cleaning and Moving
          031.2100   Foundations
               031.2110   Spread Footings
               031.2120   Pile Caps
               031.2130   Equipment Pads
          031.2200   Slabs on Grade
               031.2210   Sidewalks
               031.2220   Stairs on Grade
               031.2230   Exterior Curb and Gutter
               031.2240   Concrete Paving
               031.2250   Special Finishes
          031.2300   Columns
               031.2310   Conventionally Formed
               031.2320   Prefabricated Forms
               031.2330   Disposable Forms
          031.2400   Walls
               031.2410   Conventional Forming
                     031.2411   Straight Walls
                     031.2412   Curved Walls

Figure 13.1b

		031.2420	Gang Formed	
			031.2421	Straight Walls
			031.2422	Curved Walls
		031.2430	Special Surface Finish Formwork	
	031.2500	Elevated Slabs		
		031.2510	Flat Slabs	
		031.2520	Drop Panel Slabs	
		031.2530	Pan or Dome Slabs	
		031.2540	Flying Forms	
		031.2550	Metal Deck Forming Systems	
032.0000	Concrete Reinforcing			
	032.1000	Standard Reinforcing		
	032.2000	Masonry Reinforcing		
	032.3000	Epoxy Coated Reinforcing		
	032.4000	Prestress Tendon Reinforcing		
	032.5000	Welded Wire Fabric		
	032.6000			
	032.7000			
	032.8000	Unloading & Sorting		
	032.9000	Reinforcing Accessories		
033.0000	Concrete Placing and Finishing			
	033.1000	Concrete Materials		
		033.1100	Standard Concretes	
		033.1200	Lighweight Concretes	
	033.2000	Placing Labor and Equipment		
		033.2100	Foundations	
		033.2200	Slabs on Grade	
		033.2300	Columns	
		033.2400	Walls	
		033.2500	Elevated Slabs	
		033.2600	Exterior Curb and Gutter	
		033.2700	Concrete Paving	
	033.3000	Finishing		
		033.3100	Rubbing Wall Surfaces	
		033.3200	Float and Trowel Finishing	
		033.3300	Curing	
		033.3400	Sandblasting/Bush Hammer	
		033.3500	Surface Toppings	

Figure 13.1b Cont.

Re-cast Estimate into Budget

In reworking the information in the estimate, the project or cost engineer (or whoever is assigned the task) must ensure that the budget does several things. First, it must break the job down into groups of work which are recognizable and meaningful entities. Fortunately, a company with a working cost code system probably already has taken care of this organization process with a system that has evolved over time and proven workable. If not, the project manager must ensure that the categories established are neither too broad nor too narrow to be useful. Second, each category must contain two elements: 1) a quantity of work to be done; and 2) the resources to cover that work. An example of re-casting an estimate into a budget is shown in Figure 13.4, which contains the budget information for a portion of the sample hotel

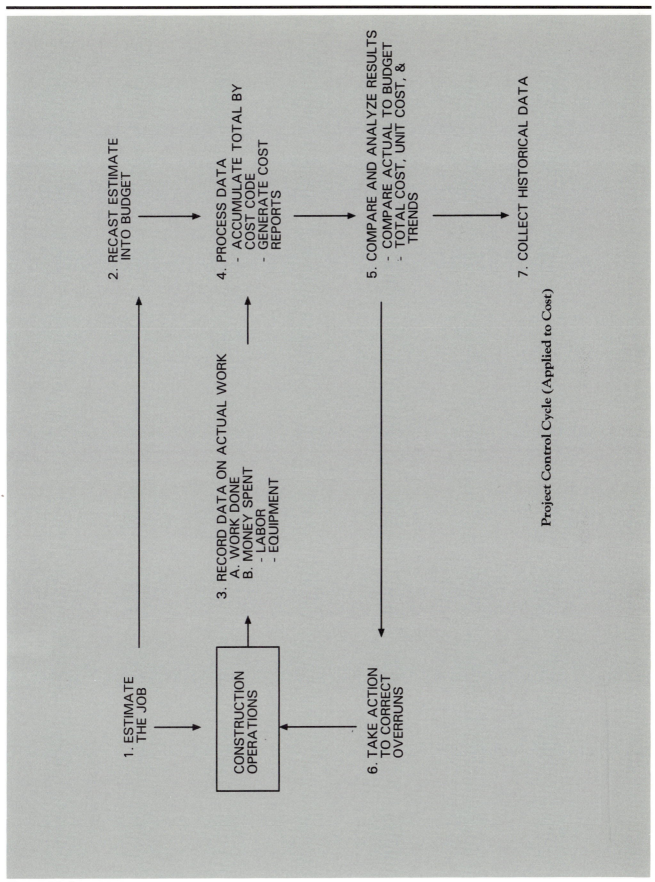

Project Control Cycle (Applied to Cost)

Figure 13.2

Means Forms
CONSOLIDATED ESTIMATE

PROJECT Sample - Extension of est. for Foundation + Waffle Slab

CLASSIFICATION

ARCHITECT

TAKE OFF BY DP QUANTITIES BY DP PRICES BY DP EXTENSIONS BY DP

SHEET NO. 1

ESTIMATE NO.

DATE

CHECKED BY

NO.	DESCRIPTION	mH/Unit	Total mH	QUANTITIES	UNIT	MATERIAL UNIT	MATERIAL UNIT COST	MATERIAL TOTAL	LABOR UNIT COST	LABOR TOTAL	EQUIP. UNIT COST	EQUIP. TOTAL	TOTAL UNIT COST	TOTAL
	Excavation	.016	145	9072	CY		1.00	9072	.31	2812	.49	4445		16329
		.08	13	158	CY		1.00	158	1.52	240	1.29	204		602
	Formwork- Wall Footing	.006	129	1962	SF		.28	549	1.32	2590	.04	78		3217
	Column Footing	.077	122	1578	SF		.35	552	1.55	2446	.05	79		3077
	Wall Forms	.122	1363	11172	SF		.60	6703	2.52	28153	.07	782		35638
	Column Forms	.136	1004	7384	SF		.53	3914	2.73	20158	.09	665		24737
	Waffle Slabs	.097	2689	27720	SF		.91	25225	2.01	55717	.05	1386		82328
	Reinforcing- Wall Footing	8.89	7	.8	Ton		515.⁴	412.00	1200.⁰⁰	1601.00	1.00	1		513
	Column Footing	8.89	9	1	Ton		515.⁰⁰	515	200.⁰⁰	2001.00	1.00	1		716
	Wall Forms	10.67	28	2.6	Tons		535.⁰⁰	1391	240.⁰⁰	624	.01	3		2018
	Column Forms	13.9	285	20.5	Tons		525.⁵⁰	10763	315.⁰⁰	6458	1.00	21		17242
	Waffle Slabs	11.03	443	40.2	Tons		540.⁹⁰	21708	250.⁰⁰	10050	1.00	40		31798
	Unloading	.56	36	651	Tons		1.00	651	12.40	8074	4.39	286		11158
	Crane	.415	27	651	Tons		1.00	651	13.45	8716	4.79	311		11052
	Place Concrete- Wall Footing	.4	28	71	CY		53.61	3811	7.00	497	.43	31		4339
	Column Footing	.813	30	34	CY		53.61	1825	15.25	519	.94	32		2376
	Wall Forms	.753	99	132	CY		53.67	7084	9.35	1234	.57	75		8393
	Column Forms	1.07	97	91	CY		53.67	4884	19.05	1734	8.20	746		7364
	Waffle Forms	.427	505	995	CY		53.61	53402	7.06	7564	3.27	3254		64218
	Finishing	110'	264	26124	SF		1.00	26124	.22	5879	.01	1069		33076
			2714					178820		148716		13509		341047

Figure 13.3

project waffle slab structure, specifically the labor for the columns and the waffle slabs.

Notice that the estimate summary (Figure 13.3) shows the cost of column formwork to be .136 man-hours per square foot of contact area, with a bare labor cost of $2.73 per square foot. The estimator, in the interest of bidding efficiency, used an aggregate number from the 1988 edition of *Means Building Construction Cost Data,* line No. 031–142–6150 (Figure 13.5). The conversion of the quantity is simple: there are 7384 square feet of contact area (SFCA) column formwork; this becomes the budget quantity.

The labor estimate figure must be converted in order to fit the budget format, and this is where the judgment calls come in. The budget breaks the formwork down into *erecting* and *stripping.* Whoever is re-casting the estimate into a budget has to alter the numbers while still retaining the integrity of the estimate. The answer in this case is to divide the single aggregate estimate numbers proportionately. Looking at *Means Building Construction Cost Data,* Circle Reference Number 33 (Figure 13.6a & b), it can be seen that *erecting column formwork* takes .098 man-hours per square foot (9.8 M.H. per 100 S.F.), and *stripping* requires .012 per square foot. This is a roughly 90/10 split. This ratio can be applied to the estimate numbers of .136 man-hours and $2.73 per square foot.

Not all work quantities can be used as they are given in the estimate. Labor figures must be further subdivided. It is much more likely that each estimate-to-budget conversion will involve a variety of these situations, going both ways. The estimate must be revised, point by point, and judgment applied to make the conversion. Ideally, all estimators and field managers should work from the same set of cost code numbers. Unfortunately, such is not the case at the present time, although some companies are developing systems to solve this problem.

Record Data on Actual Work

As the project is proceeding, the project management staff must collect data as the resources are expended to accomplish the work. Material and subcontractor payments are not normally difficult to control.

While it is important to ensure against overuse of material and overpayments to subcontractors, it is not necessary to treat these issues with the same aggressiveness as should be directed toward careful monitoring of labor and

Concrete Budget Calculations					
Item	Units	U/L	T/L	U/MH	T/MH
fc	7384	2.45	18091	.122	901
rc	20.5	327.40	6712	14.460	296
pc	91	19.05	1734	1.070	97
sc	7384	.27	1994	.014	103
					0
fws	27720	1.71	47401	.082	2273
rws	40.2	262.40	10548	11.590	466
pws	995	7.60	7562	.427	425
sws	27720	.30	8316	.015	416
			73828		4978

Figure 13.4

031 100	Struct C.I.P. Formwork	CREW	DAILY OUTPUT	MAN-HOURS	UNIT	BARE COSTS				TOTAL INCL O&P	
						MAT.	LABOR	EQUIP.	TOTAL		
142 1500	Round fiber tube, 1 use, 8" diameter	C-1	155	.206	L.F.	1.75	4.14	.13	6.02	8.20	**142**
1550	10" diameter		155	.206		2.50	4.14	.13	6.77	9.05	
1600	12" diameter		150	.213		3	4.27	.14	7.41	9.80	
1650	14" diameter		145	.221		3.85	4.42	.14	8.41	10.95	
1700	16" diameter		140	.229		5	4.58	.15	9.73	12.45	
1750	20" diameter		135	.237		8	4.75	.15	12.90	16	
1800	24" diameter		130	.246		10	4.93	.16	15.09	18.50	
1850	30" diameter		125	.256		15.50	5.15	.16	20.81	25	
1900	36" diameter		115	.278		19	5.60	.18	24.78	29	
1950	42" diameter		100	.320		36	6.40	.20	42.60	49	
2000	48" diameter	↓	85	.376		48	7.55	.24	55.79	64	
2200	For seamless type, add					15%					
3000	Round, steel, 4 use per mo., rent, regular duty, 12" diam.	C-1	145	.221		2	4.42	.14	6.56	8.95	
3050	16" diameter		125	.256		2.15	5.15	.16	7.46	10.15	
3100	Heavy duty, 20" diameter		105	.305		2.30	6.10	.19	8.59	11.80	
3150	24" diameter		85	.376		2.50	7.55	.24	10.29	14.20	
3200	30" diameter		70	.457		2.75	9.15	.29	12.19	16.95	
3250	36" diameter		60	.533		3	10.70	.34	14.04	19.55	
3300	48" diameter		50	.640		4.50	12.80	.41	17.71	24	
3350	60" diameter		45	.711	↓	5.60	14.25	.45	20.30	28	
4000	Column capitals, 4 use per mo., 24" col, 4' cap diameter		12	2.670	Ea.	13.75	53	1.70	68.45	96	
4050	5' cap diameter		11	2.910		13.45	58	1.85	73.30	105	
4100	6' cap diameter		10	3.200		17.30	64	2.04	83.34	115	
4150	7' cap diameter	↓	9	3.560		19	71	2.27	92.27	130	
4500	For second and succeeding months, deduct				↓	50%					
4600											
5000	Plywood, 8" x 8" columns, 1 use ㉜	C-1	165	.194	SFCA	1.70	3.89	.12	5.71	7.80	
5050	2 use		195	.164		.95	3.29	.10	4.34	6.05	
5100	3 use ㉝		210	.152		.70	3.05	.10	3.85	5.40	
5150	4 use		215	.149		.60	2.98	.09	3.67	5.20	
5500	12" x 12" plywood columns, 1 use		180	.178		1.65	3.56	.11	5.32	7.25	
5550	2 use		210	.152		.94	3.05	.10	4.09	5.65	
5600	3 use		220	.145		.67	2.91	.09	3.67	5.15	
5650	4 use		225	.142		.55	2.85	.09	3.49	4.94	
6000	16" x 16" plywood columns, 1 use		185	.173		1.60	3.47	.11	5.18	7.05	
6050	2 use		215	.149		.85	2.98	.09	3.92	5.50	
6100	3 use		230	.139		.64	2.79	.09	3.52	4.94	
6150	4 use		235	.136		.53	2.73	.09	3.35	4.73	
6500	24" x 24" plywood columns, 1 use		190	.168		1.41	3.37	.11	4.89	6.70	
6550	2 use		216	.148		.79	2.97	.09	3.85	5.40	
6600	3 use		230	.139		.58	2.79	.09	3.46	4.88	
6650	4 use		238	.134		.48	2.69	.09	3.26	4.63	
7000	36" x 36" plywood columns, 1 use		200	.160		1.70	3.21	.10	5.01	6.75	
7050	2 use		230	.139		.90	2.79	.09	3.78	5.25	
7100	3 use		245	.131		.65	2.62	.08	3.35	4.70	
7150	4 use		250	.128		.52	2.56	.08	3.16	4.47	
7500	Steel framed plywood, 4 use per mo., rent, 8" x 8"		290	.110		.56	2.21	.07	2.84	3.98	
7550	10" x 10"		300	.107		.44	2.14	.07	2.65	3.73	
7600	12" x 12"		310	.103		.37	2.07	.07	2.51	3.56	
7650	16" x 16"		335	.096		.35	1.91	.06	2.32	3.30	
7700	20" x 20"		350	.091		.31	1.83	.06	2.20	3.13	
7750	24" x 24"		365	.088		.26	1.76	.06	2.08	2.96	
146 0010	FORMS IN PLACE, CULVERT 5' to 8' square or rectangular, 1 use		170	.188		1.52	3.77	.12	5.41	7.40	**146**
0050	2 use		180	.178		.91	3.56	.11	4.58	6.40	
0100	3 use		190	.168		.71	3.37	.11	4.19	5.90	
0150	4 use ㉝	↓	200	.160	↓	.60	3.21	.10	3.91	5.55	
150 0010	FORMS IN PLACE, ELEVATED SLABS										**150**
0050	See also corrugated form deck, division 053-104										

For expanded coverage of these items see *Means Concrete Cost Data 1988*

73

Figure 13.5

CIRCLE REFERENCE NUMBERS

(33) Forms in Place (cont.)

COLUMNS, 24" x 24" (Line 142-6500)	First Use			Reuse		
	Quantities	Material	Installation	Quantities	Material	Installation
5/8" exterior plyform @ $650 per M.S.F.	120 S.F.	$ 78.00		12.0 S.F.	$ 7.80	
Lumber @ $340 per M.B.F.	125 B.F.	42.50		12.5 B.F.	4.25	
Clamps, chamfer strips and accessories	Allow	20.80		Allow	4.60	
Make up, crew C-1 at $20.66 per man-hour	5.8 M.H.		$119.85	1.0 M.H.		$ 20.65
Erect and strip	9.8 M.H.		202.45	9.8 M.H.		202.45
Clean and move	1.2 M.H.		24.80	1.2 M.H.		24.80
Total per 100 S.F.C.A.	16.8 M.H.	$141.30	$347.10	12.0 M.H.	$16.65	$247.90

FLAT SLAB WITH DROP PANELS (Line 150-2000)	First Use			Reuse		
	Quantities	Material	Installation	Quantities	Material	Installation
5/8" exterior plyform @ $650 per M.S.F.	115 S.F.	$ 74.75		11.5 S.F.	$ 7.50	
Lumber @ $340 per M.B.F.	210 B.F.	71.40		21 B.F.	7.15	
Accessories, incl. adjustable shores	Allow	22.40		Allow	22.40	
Make up, crew C-2 at $21.31 per man-hour	3.5 M.H.		$ 74.60	1.0 M.H.		$ 21.30
Erect and strip	6.0 M.H.		127.85	6.0 M.H.		127.85
Clean and move	1.2 M.H.		25.55	1.2 M.H.		25.55
Total per 100 S.F.C.A.	10.7 M.H.	$168.55	$228.00	8.2 M.H.	$37.05	$174.70

Drop panels included but column caps figure with columns.

FOOTINGS, SPREAD Line (158-5000)	First Use			Reuse		
	Quantities	Material	Installation	Quantities	Material	Installation
Lumber @ $340 per M.B.F.	260 B.F.	$ 88.40		26 B.F.	$ 8.85	
Accessories	Allow	6.50		Allow	6.50	
Make up, crew C-1 at $20.66 per man-hour	4.7 M.H.		$ 97.10	1.0 M.H.		$ 20.65
Erect and strip	4.2 M.H.		86.75	4.2 M.H.		86.75
Clean and move	1.6 M.H.		33.05	1.6 M.H.		33.05
Total per 100 S.F.C.A.	10.5 M.H.	$ 94.90	$216.90	6.8 M.H.	$15.35	$140.45

FOUNDATION WALL, 8' High (Line 182-2000)	First Use			Reuse		
	Quantities	Material	Installation	Quantities	Material	Installation
5/8" exterior plyform @ $650 per M.S.F.	110 S.F.	$ 71.50		11.0 S.F.	$ 7.15	
Lumber @ $340 per M.B.F.	140 B.F.	47.60		14.0 B.F.	4.75	
Accessories	Allow	16.40		Allow	16.40	
Make up, crew C-2 at $21.31 per man-hour	5.0 M.H.		$106.55	1.0 M.H.		$ 21.30
Erect and strip	6.5 M.H.		138.50	6.5 M.H.		138.50
Clean and move	1.5 M.H.		31.95	1.5 M.H.		31.95
Total per 100 S.F.C.A.	13.0 M.H.	$135.50	$277.00	9.0 M.H.	$28.30	$191.75

PILE CAPS, Square or Rectangular (Line 158-3000)	First Use			Reuse		
	Quantities	Material	Installation	Quantities	Material	Installation
5/8" exterior plyform @ $650 per M.S.F.	110 S.F.	$ 71.50		11.0 S.F.	$ 7.15	
Lumber @ $340 per M.B.F.	160 B.F.	54.40		16.0 B.F.	5.45	
Accessories	Allow	6.50		Allow	6.50	
Make up, crew C-1 at $20.66 per man-hour	4.5 M.H.		$ 92.95	1.0 M.H.		$ 20.65
Erect and strip	5.0 M.H.		103.30	5.0 M.H.		103.30
Clean and move	1.5 M.H.		31.00	1.5 M.H.		31.00
Total per 100 S.F.C.A.	11.0 M.H.	$132.40	$227.25	7.5 M.H.	$19.10	$154.95

STAIRS, Average Run (Inclined Length x Width) (Line 174-0010)	First Use			Reuse		
	Quantities	Material	Installation	Quantities	Material	Installation
5/8" exterior plyform @ $650 per M.S.F.	110 S.F.	$ 71.50		11.0 S.F.	$ 7.15	
Lumber @ $340 per M.B.F.	425 B.F.	144.50		42.5 B.F.	14.45	
Accessories	Allow	16.50		Allow	16.50	
Build in place, crew C-2 at $21.31 per man-hour	25.0 M.H.		$532.75	25.0 M.H.		$532.75
Strip and salvage	4.0 M.H.		85.25	4.0 M.H.		85.25
Total per 100 S.F.	29.0 M.H.	$232.50	$618.00	29.0 M.H.	$38.10	$618.00

Figure 13.6a

equipment on the actual construction. The recording of expenditures for material does not take on the same urgency as that for labor and equipment, and it is often done by persons other than direct project management personnel.

Recording and tracking labor and equipment costs is the direct concern of the project manager and anyone else who is involved in controlling the production process. Specifically, these individuals must record three primary categories of information: work performed, labor expended, and equipment used.

Recording Work Performed   In order to determine how much money — or labor — should have been spent at any point in the project, it is necessary to know how much work has been accomplished. For example, if the labor budget for column formwork is $18,091.00, but not all the formwork has been erected, then it is clear that not all of the money should have been spent. The money should be spent at a rate proportional to the rate of work being done. To determine this rate, the project manager should know how much work has been accomplished.

Unfortunately, calculating the work completed can be a difficult task. Again, the judgment of the project manager (or other person calculating the values) is a key factor. There are several general approaches to the problem. The one that works best should be used for each individual item on the job. The first is *direct measurement of work*. This approach involves someone physically measuring what has been installed. This system is typically used on jobs where extensive surveying is done anyway. An example is roadwork,

33 Forms In Place (Div. 031)

This section assumes that all cuts are made with power saws, that adjustable shores are employed and that maximum use is made of commercial form ties and accessories. Bare costs are used in the table below.

BEAM AND GIRDER, INTERIOR, 12" Wide (Line 138-2000)	First Use			Reuse		
	Quantities	Material	Installation	Quantities	Material	Installation
5/8" exterior plyform at $650 per M.S.F.	115 S.F.	$ 74.75		11.5 S.F.	$ 7.50	
Lumber at $340 per M.B.F.	200 B.F.	68.00		20.0 B.F.	6.80	
Accessories, incl. adjustable shores	Allow	20.65		Allow	20.65	
Make up, crew C-2 at $21.31 per man-hour	6.4 M.H.		$136.40	1.0 M.H.		$ 21.30
Erect and strip	8.3 M.H.		176.90	8.3 M.H.		176.90
Clean and move	1.3 M.H.		27.70	1.3 M.H.		27.70
Total per 100 S.F.C.A.	16.0 M.H.	$163.40	$341.00	10.6 M.H.	$34.95	$225.90

For structural steel frame with beams encased, subtract 1.2 man-hours, and 50 M.B.F. lumber or about $40 per 100 S.F.C.A. for the first use and $26 for each reuse.

BOX CULVERT, 5' to 8' Square or Rectangular (Line 146-0010)	First Use			Reuse		
	Quantities	Material	Installation	Quantities	Material	Installation
3/4" exterior plyform at $705 per M.S.F.	110 S.F.	$ 77.55		11.0 S.F.	$ 7.75	
Lumber at $340 per M.B.F.	170 B.F.	57.80		17.0 B.F.	5.80	
Accessories	Allow	16.25		Allow	16.25	
Build in place, crew C-1 at $20.66 per man-hour	14.5 M.H.		$299.55	14.5 M.H.		$299.55
Strip and salvage	4.3 M.H.		88.85	4.3 M.H.		88.85
Total per 100 S.F.C.A.	18.8 M.H.	$151.60	$388.40	18.8 M.H.	$29.80	$388.40

Figure 13.6b

where a state surveyor typically provides control points and profiles for the contractor.

The second method uses the same quantity survey techniques used in estimating. If the project manager can see by looking at a building how far up and along a wall the brick work has gone, then the quantities complete can be calculated using the plans and a scale and calculator, and standard quantity surveying factors, such as number of bricks per square foot.

A third method is to use the recording of activities completed in the scheduling system, especially if the calculation of activity times was done using methodologies such as those described in Chapter 6. Finally, it should be noted that for many types of work it is virtually impossible to calculate work done with significant precision, and a certain amount of approximation has to be accepted. An example is *reinforcing steel.* It may be possible to work from the shop drawing, or physically count bars and then calculate lengths and multiply by bar weights per lineal foot, but by the time a cost engineer has completed the task, the reporting period would be over.

Whatever degree of precision is accepted, and whatever method is used, some attempt should be made to establish work complete. When the progress has been calculated, it should be recorded as shown in Figure 13.7. This example shows how a typical week's progress for the waffle slab work might be recorded.

Recording the Resources Expended Recording the actual expenditures of labor is done using daily or weekly labor time cards. Filling out these cards is fairly self-explanatory. Examples of each are shown in Figures 13.8 and 13.9. Each is filled out for a representative day or week for the small waffle slab illustrated in this chapter. The foreman enters the appropriate data based on the specific work performed by the crew. Absolute precision may not be possible, but if the time is recorded promptly at the end of the day, or notes made as the work progresses, a conscientious supervisor should be accurate to within the nearest one half hour.

In general, using daily time cards is a more accurate method than using weekly time sheets. The time is recorded while it is still fresh in the mind of the foreman or other person responsible for filling out the card. Whichever method is used, the responsibility for recording and checking the data should be firmly established in the job site procedures. Cards should be checked daily for accuracy of work categories and hours recorded, and for human error.

The time card system should be designed for ease of use by the people in the field who must fill them out. For example, the time card can be computer generated with as much information as possible already recorded, such as job number, date, etc. Many companies generate time cards in advance with the employee's social security number, job assignment, and other basic information. These cards are then mailed to the site for use. Such measures for reducing administrative workload can significantly improve the accuracy of the system, as well as the degree to which it is accepted in the field.

Finally, the job labor cost system should be tied in with the payroll system of the company; the time card should serve both needs. Since the same information must be collected for both functions, the savings in overhead from using one recording form can be considerable.

Collecting costs on equipment can be done using the same form of cards. Again, daily records are better than weekly ones, for the same reasons. It should be noted that it is often not possible to assign equipment costs to precise categories of work. A case in point would be a large crane used to lift

Means Forms

**PERCENTAGE
COMPLETE ANALYSIS**

PROJECT **Sample Hotel**

PAGE 1

DATE **April 18, 1988**

ARCHITECT

BY

FROM — TO —

NO.	DESCRIPTION	ACTUAL OR ESTIMATED	TOTAL PROJECT	THIS PERIOD QUANTITY	%	QUANTITY	PERCENT TOTAL TO DATE 10	20	30	40	50	60	70	80	90	100
20010	FRP COLS E-BSMT TO mA	ACTUAL	2766	O		2766										X
		ESTIMATED	SFCA													
20030	FRP WAFF SLABS-nE MAIN	ACTUAL	5080	O		5080										X
		ESTIMATED														
20020	FRPS COLS w-BSMT TO mn	ACTUAL	745	O		745										
		ESTIMATED														
20040		ACTUAL														
		ESTIMATED														
20040	FRP WAFFLE SLABS-SE MAIN	ACTUAL	5080	3150		3150						X				
		ESTIMATED														
		ACTUAL														
		ESTIMATED														
		ACTUAL														
		ESTIMATED														
		ACTUAL														
		ESTIMATED														
		ACTUAL														
		ESTIMATED														
		ACTUAL														
		ESTIMATED														
		ACTUAL														
		ESTIMATED														
		ACTUAL														
		ESTIMATED														
		ACTUAL														
		ESTIMATED														
		ACTUAL														
		ESTIMATED														
		ACTUAL														
		ESTIMATED														
		ACTUAL														
		ESTIMATED														
		ACTUAL														
		ESTIMATED														
		ACTUAL														
		ESTIMATED														
		ACTUAL														
		ESTIMATED														

Figure 13.7

formwork and reinforcing into place, but also used on occasion to move interior materials to the upper floors of a building. In this case, it may be better to record the crane costs in a general job cost category, making sure that the budget is set up to reflect that fact.

Process Data

This step is not the responsibility of the project manager. It involves feeding the data into the accounting system, by hand or machine. The accumulations and totals are developed, and the reports printed for the project manager's use. Only in the case of very large projects is this process carried out on the job site.

The job cost system is part of the overall financial accounting system of a construction company. Every construction company must have the standard parts of an accounting system — general ledger, accounts payable, accounts receivable, etc. These are derived to some degree from the job cost system since all transactions begin with construction work done. However, these parts of the accounting system only affect the project manager in a peripheral way. The key to job site success, and the success of the company, rests with the control of production costs, and this must occur at the job site.

Another issue is whether the job cost system should be kept by hand or by computer. The fact is, there are almost no circumstances where hand systems make sense. The few cases they might be better are those in which all work is subcontracted and no significant company labor is used. These situations do not reflect true job cost control, but rather issues of cash flow management. Further, the quality of construction accounting cost control systems has

			DESCRIPTION OF WORK						TOTALS		RATES		OUTPUT	
									REG-ULAR	OVER-TIME	REG-ULAR	OVER-TIME		
NO.	NAME													
	J. Kubasok	HOURS	8						8		21.20		SEE	
		UNITS											PROGRESS	
	J. Borland	HOURS	4	4					8		21.20		REPORT	
		UNITS												
	n. morris	HOURS	8						8		21.20			
		UNITS												
	J. Emmack	HOURS	4	4					8		21.20			
		UNITS												
		HOURS												
		UNITS												
		HOURS												
		UNITS												
		HOURS												
		UNITS												

Means Forms
DAILY TIME SHEET
PROJECT Sample Hotel
FOREMAN Rodger
WEATHER CONDITIONS Fair - cool
TEMPERATURE 58°F
DATE April 1, 1988
SHEET NO. 1

Figure 13.8

greatly improved in the recent past. Some very good systems are available for reasonable prices. Besides, a microcomputer used for a small accounting system can also serve many other purposes, further reducing the cost of the system.

It should be noted, however, that there are vendors of computer systems who do not know the construction business and try to sell systems adapted from other fields. These systems may not work at best, and may cause a lot of damage at worst. As is true of any purchase of a major capital item, the contractor must do the necessary investigation to ensure that the system fits a specific situation.

Compare and Analyze the Results

Once the weekly summary reports have been received, the project manager must analyze the information presented to determine the status of the job at the end of the reporting period. The information generated should contain several elements, all of which are important to determine where action needs to be taken. If the system does not include all of the elements, further analysis of raw data, or of the information contained in the reports generated may also be necessary.

Information Needed   First, the information must be timely. After the labor and equipment cost data has been collected and turned into the accounting department, it should be processed and returned to the field within a few days at most. Promptness is absolutely essential if control is to be maintained over

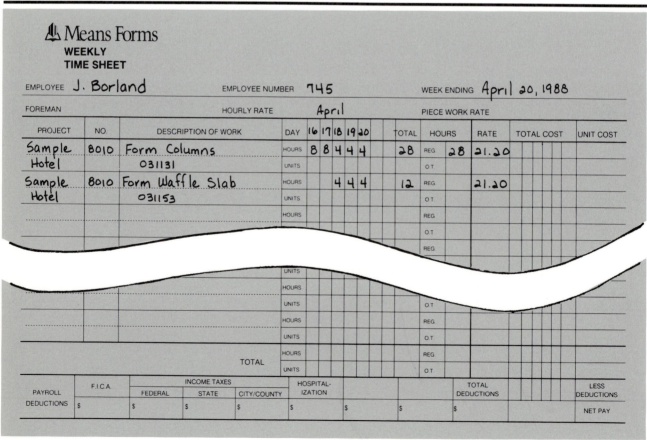

Figure 13.9

the production process. Labor and equipment costs can quickly overrun due to the volatile nature of the construction processes. Fast action is essential.

Second, there are several types of information that have proven useful in identifying the source and extent of overruns. Figure 13.10 is an example of a fairly comprehensive type of labor cost summary containing most of the types of information needed. The key elements to look at are:

1. Unit costs relative to budget
2. Total costs relative to budget (particularly relative magnitude of overruns)
3. Trends in costs

Reviewing the report for the above elements — section by section — reveals the information of value. The first two sections (Columns 1–8) remain constant throughout the life of the job, barring changes in scope of work or price. Each item is identified by cost code. The budget information is presented for the work to be done, in terms of budgeted man-hours and budgeted dollars. Man-hours and dollars are presented in both *totals* and *cost per unit of work.*

The third section (Columns 9–11) contains the information about work performed during the period covered by the report, and the total work done to date on the job. These figures are also converted to a percent complete for each category of work. The fourth section (Columns 12–13) provides information on budgeted man-hours and costs to date. These figures represent the maximum amount that should have been spent accomplishing the work reported and recorded in the previous section. The budgeted expenditures are determined by multiplying the percent complete in Column 11 times the original budgeted amounts in Columns 6 and 8.

The last two sections provide the project manager with information on accumulated costs and comparisons with the budgeted amounts. Section 5 contains information on total man-hours and dollars; section 6 offers the same information on unit man-hours and dollars. Note that both sections provide information about the period just covered (*this week* columns), and the total progress so far (*to date* columns). Also presented are columns that "summarize" the sections by giving over/under figures and percent deviation figures.

What the Information Reveals   A review of the report in Figure 13.10 informs the project manager on a number of aspects of the job. From the total for Column 18, it can be seen that this segment is $90 under budget at this point, which is about 20% complete (approximately $20,000 spent out of a budgeted amount of $102,000). The report also shows, however, that several areas of the job are significantly over budget, others are significantly under. This is often very typical of jobs which appear to be on target overall.

Looking at the detail in Columns 18 and 23 reveals the specifics. Forming is both over and under budget. Columns are over by 6%; waffle slabs are under by approximately the same amount. Reinforcing shows the same kind of variance. This particular set of data does not tell the project manager if forming and reinforcing are general problems. Reviewing the week's performance to date reveals significant data, however. In this case, it shows that the cost of forming columns is going down; the unit man-hours and unit dollars are both lower for the week just ended than they are to date.

Example Weekly Labor Cost Report
Sample Hotel Project

1	2	3	Original Budget					Work Done			Budget to Date		Total Cost/Total MH					Unit Cost/Unit Man-hours				
			4	5	6	7	8	9	10	11	12	13	14	15	16	17	18	19	20	21	22	23
Item	Cost Code	Units	#	MH/Unit	Total MH	$/Unit	Total $	This Week	To Date	% Compl	MH	$	MH This Week	MH To Date	$ This Week	$ To Date	Over or Under	Unit This Week	MH This Week	$ This Week	$ To Date	% Dev.
Form Columns		SFCA	7384	.122	901	2.45	18091	1936	5447	73.8	665	13345	184	695	3754	14178	833	.095	.128	1.94	2.60	106.2
Reinforce Columns		TNS	20.5	14.460	296	327.40	6712	5.4	15.2	74.1	220	4976	67	205	1518	4643	−333	12.407	13.487	281.03	305.48	93.3
Place Columns		CY	91	1.070	97	19.05	1734	24	67	73.6	72	1276	34	90	607	1606	329	1.417	1.343	25.27	23.96	125.8
Strip Columns		SFCA	7384	.014	103	.27	1994	1936	5447	73.8	76	1471	27	76	527	1482	11	.014	.014	.27	.27	100.8
Form Waffle Slabs		SFCA	27720	.082	2273	1.71	47401	3810	8890	32.1	729	15202	324	702	6610	14321	−881	.085	.079	1.73	1.61	94.2
Reinf. Waffle Slabs		TNS	40.2	11.590	466	262.40	10548	5.6	12.9	32.1	150	3385	70	174	1586	3941	556	12.500	13.488	283.13	305.51	116.4
Place Waffle Slabs		CY	995	.427	425	7.60	7562	0	185	18.6	79	1406	0	92	0	1641	235		.497		8.87	116.7
Strip Waffle Slabs		SFCA	27720	.015	416	.30	8316	0	0	.0	0	0	0	0	0	0	0					
Totals:					4977		102,358				957	19993	394	968	8195	19903	−90					

Figure 13.10

Reinforcing columns, while still below budget, shows a decreasing cost trend. *Forming waffle slabs* is under budget but shows an increasing cost trend. Looking further, it is apparent that *placing concrete* is a general problem. Both *column and waffle slab placing* are significantly over budget, one by 25% and the other by 16%. *Placing columns* is over in unit man-hours and unit dollars and both are on an upward trend. These trends are all revealed by the unit costs for the week and to date.

Take Action to Correct Overruns

While labor and equipment cost reports identify the location of the problem, they do not reveal the cause of the problem. The question is: which areas are a concern and ought to be investigated? Sometimes the answer is clear. In this case, the first and most obvious is the task of placing concrete. This activity is in trouble across the board. Both unit costs are over budget by large amounts, and the project manager might be well advised to consider subcontracting the remainder of the building if a good price can be found.

The next areas of concern are not so easy to identify. The project manager must examine each task in detail and decide where his intervention will be the most productive. To make these decisions, he must consider the following factors. What is the total amount over budget? What is the trend of the cost over the last several reporting periods? Is the activity too far along to be "saved"? These may all be contributing factors and the answers may not be clear cut.

Checking each cost code in turn, the following becomes apparent. Forming columns is way over budget, but is on a downward trend and only 25% of the work remains. Remedial action, therefore, would have a limited effect. Reinforcing columns is under budget, on a downward trend, and almost complete. A profit will be made on this work and the activity can be left alone. Placement of columns is clearly in need of attention. *Stripping columns* is not a problem.

Forming waffle slabs is under budget, but the cost trend is increasing, and 65% of the work remains. This case calls for preventive action to ensure that the 6% margin established thus far does not deteriorate. *Reinforcing waffle slabs* is another area that clearly needs attention. It is significantly over budget though the trend is down, and much of the work remains.

Once the areas of concern are identified, follow-up is essential to a successful cost control process. The technical causes of the overruns must be determined, and this can be done only by talking to people in the field. Again, the manner in which these individuals are approached about a problem can have a significant outcome on its resolution. If they perceive any sort of a threat to their jobs, they may provide a biased version of the facts. Such distortion only complicates the task of bringing the costs back into line. The project manager should instead approach the superintendents with an attitude of joint problem solving. If he conveys a willingness to work with them, their responses should be much more positive and productive.

Other Cost Control Issues

Equipment Cost Records and Reports

The task of recording, collecting, and reporting equipment costs is very similar to the process applied to labor. The time cards, report formats, and information that the project manager must obtain from the reports are essentially the same as those covered in the section on labor cost control.

There are some special considerations involved in recording and reporting equipment costs. First is the issue of hourly rates to be charged to the job for various pieces of equipment. In the case of labor, the project manager has very little input. The requirements of the contract, especially on any kind of public work, and the hiring policies of the company (union or non-union) govern the hourly wage to be paid to employees. Also, the calculation of total pay is handled through the payroll system. This is not the case with equipment. Sometimes the project manager has the authority to choose where to obtain equipment based on cost to the job, and the resulting cost must be properly recorded in the project cost system.

Three methods for determining the cost to be charged to the job are commonly used in the industry. The first is to charge the actual cost. This is unusual, however, since the issue of how to determine cost is clouded by the issue of how to charge depreciation. This method tends to be used only on very large jobs where the entire useful life of an individual piece of equipment is to be used on the one specific project to which it is charged.

The second and most common method is to keep all cost records for owning, running, and maintaining the individual pieces of equipment in a cost account separate from the cost of the overall job. The total costs of owning the equipment can then be calculated. An hourly rate is determined and that rate charged against the job for each hour it is used. The hourly rate for equipment takes into account all depreciation, interest, fuel, repairs, and other expenses. At the end of the fiscal year, the *actual* hourly cost of the equipment can be determined. The company then knows if the equipment cost more per hour than it "earned" from being charged to jobs, and can make an intelligent decision about whether or not it is good company policy to own the individual equipment or to rent or lease from outside sources.

Another variation of the separate account method is charging the project for company equipment. The rate charged is based on the local market rate for that type of equipment, and calculated as cost per hour. If the equipment "earnings" from the job do not exceed the actual costs for the equipment, then clearly, the company should instruct the project manager to rent from outside sources.

The third method covers equipment rented from outside sources. In this case, the equipment is charged at the actual rate which has to be paid to the vendor.

Whatever method is used to determine the hourly rate or cost, the project manager should try to ensure that equipment costs are charged to the type of work for which the equipment is actually used. Some companies take the view that all equipment should be charged to one central account on the project, but this is not good practice. As is the case with labor, equipment cost is a volatile item, easily subject to overruns, and should be controlled by means of specific cost codes or identification of types of work. If the project manager charges equipment costs to a single account on the job, he gives up any possibility of identifying *where* equipment costs are out of control, should such an event occur. Small tools (saws, drills, temporary electrical distribution equipment, etc.) are the only exception to this rule, and need not be charged to each specific work activity for which they are used.

Collect Historical Data

While complete coverage of estimating is beyond the scope of this book, it should be noted that the best single source of acccurate data for estimating future jobs is the company's own cost and productivity records. These records reflect the company's actual performance and, as such, are the best predictor of future performance.

The problem with using company data is the fact that very few firms have a formal feedback system to consult historical costs for the development of estimates. Even less likely is access to a computer system designed for easy retrieval of this historical data. More often than not, the "system" consists of someone digging through past job reports to find unit costs or productivities to fit the specific job being bid. Even when a similar job is found, the cost variation can often be substantial. The estimator must apply judgment in adjusting the unit cost to make an accurate prediction for future work.

14

Summary and Conclusions

Summary and Conclusions

To perform effectively, the project manager must work within a framework of proven management principles. Many construction firms have discovered that the planning and control process works best when it is modeled on the *Project Control Cycle* concept presented in Chapter 2. This cycle is an extension of the basic feedback cycle which states that the project manager must always have in mind the following:

1. Establishing a reasonable plan of action for building the project;
2. Monitoring progress toward that goal; and
3. If the goal is not being met, taking well-thought-out and assertive action to correct the situation.

Applying these principles will improve the project's chances for success, whether the techniques used are simple or sophisticated. The project manager need not plan or execute the project with absolute precision; the important thing is to initially plan the project, track it, and then take action to correct any problem areas.

The overall control process begins with pre-construction planning. It is important that the project manager and the entire project team be fully informed — from the start — on the project and its unique requirements. This is accomplished through thorough reviews of the contract documents, drawings, specifications, and estimates, and through interviews with estimators and others who have worked on the project.

Once the initial pre-planning has been done, the control process breaks down into two separate and distinct tasks:

1. Scheduling, or control of the project's progress toward a completion date, and
2. Cost control, or control of expenditures in order to meet budget and profit targets.

The first of these two functions, time control, is effectively accomplished using what is known as the *Critical Path Method, or CPM*. This technique has proven to be extremely well suited to construction company needs. With the computer systems available today, CPM has also become a very cost effective method.

Step 1 of the scheduling process is the division of the project's construction work into smaller, manageable tasks, known as *activities*. Step 2 involves listing these activities in the order in which they will be performed, using a format known as a *logic diagram*. Once the logic has been established, the project manager can carry out Step 3, determining predicted times for each individual activity by applying established production rates to the quantities of work in the activities. When the individual tasks have been accounted for, Step 4, an analysis of the overall job, can be performed. The result is a scheduled starting and finishing time for each individual activity within the overall project schedule and completion time for the project as a whole. These four steps lead to a plan of action and a set of times that the project manager can use in directing the labor, material, and equipment to complete the project within the allotted contract time.

In addition to the actual construction work, it is vitally important that the project manager also incorporate and control the procurement process. This separate schedule of administrative tasks must then be integrated with the construction schedule to ensure timely delivery of the material essential to construction.

While the schedule is being developed, the project manager should ensure that various subcontractors and company managers are properly informed and kept up-to-date. Most construction schedules should be organized in such a way that different levels of information can be reported to the various levels of management. Lower level managers need to see more day-to-day detail, and higher level managers must see the overview. Without this selective reporting capability, it is all too easy for managers to become overwhelmed (or distracted) with extra and unnecessary detail.

When the planning and scheduling have been accomplished, the actual construction must be monitored closely. As the work takes place, the project manager must see to it that all the activities are tracked and the status of the work reported to the proper managers. This part of the process is known as *updating*. The updating task becomes part of the CPM scheduling process because it provides information helpful in predicting future activity durations. This update information is vital to the project manager in carrying out the last step, evaluating the progress of the project as a whole.

In this last phase, the project manager uses the actual schedule results to determine the status of the project relative to the original goals. If the project is behind schedule, then the project manager must take aggressive, positive action in order to correct the situation. This is one of the project manager's most essential responsibilities, and reflects the leadership abilities required of this position. It is this *proactive* rather than *reactive* approach that makes a building project successful.

In addition to controlling the project time progress, the CPM method can be used to more effectively manage resources such as labor and equipment, thereby ensuring a smooth, even flow of work. These resource management techniques are also used for the difficult task of reducing a project's total construction time. The CPM technique makes it easier to decide when and where to shorten project time most effectively and economically. Orderly and efficient use of resources is a very difficult part of the project manager's job, but the shrewd application of CPM-based techniques can make his decisions much more accurate, thereby increasing the project's profit.

The second of the two major project control functions is cost control. From the project manager's standpoint, cost control is somewhat simpler than time control in that many of the procedures are already established as part of the company's overall accounting system. Cost control is, however, no less vital to the project and the construction company, and may take precedence over time control in many instances. The basic idea of cost control is similar to scheduling, i.e., setting targets (in this case, unit and total costs) for every part of the project, and then monitoring results and making corrections where necessary.

Once the budget is established, the project manager must ensure that the actual costs are accurately recorded on labor and equipment time cards, and then processed into the cost system to generate cost performance reports. He must then interpret these reports, comparing unit and total cost expenditures to budgeted amounts, and then determining where and why cost overruns are occurring, if at all. Again, the project manager must take on a leadership role, aggressively following up with the proper remedial action to ensure that the causes of overruns are halted, while maintaining efficient construction procedures.

The project manager's key responsibility is to *plan, monitor, and correct where necessary.* It is an established "feedback loop" that provides the project manager with vital information on which to act. With early knowledge of deviations from a sound original plan, the project manager is in a position to truly control rather than be controlled by events.

Appendix

Table of
Contents

Appendix A
Sample Project Estimate   235

Appendix B
Sample Logic Diagrams and Computer Reports   241

Appendix C
C.S.I. Masterformat Divisions   267

```
=================================================================================================

NO. 1988-100              BURDENED ITEMIZED JOB REPORT                           PAGE   1
-------------------------------------------------------------------------------------------------

PROJECT    : SAMPLE HOTEL PROJECT              LOCATION  : KINGSTON, MASSACHUSETTS
ARCH/ENGR  : HOWARD, FINE & HOWARD, INC.       OWNER     : XYZ DEVELOPMENT
QUANTITIES BY: PLM                             ENTERED BY: PLM

=================================================================================================

                                    2 SITEWORK

-------------------------------------------------------------------------------------------------

 2.2 EARTHWORK
-------------------------------------------------------------------------------------------------

DESCRIPTION                       CREW  QUANTITY  UNIT   M/H    MATERIAL   LABOR    EQUIP   TOTAL    SUB
LINE NO.

-------------------------------------------------------------------------------------------------

XCAVTG BULK FRNTENDLOADR TRCKMNTD 2.5CY 95CY/HR
  022 238 1250 00  M               B100  9072.00  C.Y.   0.016    0.00     0.46     0.54     1.00
  -                      100% 100%                      143.24    0.00  4183.15  4902.94  9086.09


XCVTNG TRNCH/FTG 4 W 1/2-1SLOP DMPSNDYLOM 4 D 1/2CYTRCTBKHO 225LF/DAY
  022 254 0080 00  M               B11M   158.00  C.Y.   0.080    0.00     2.24     1.41     3.66
  -                      100% 100%                       12.64    0.00   354.49   223.51   578.00


=================================================================================================
                                  SUB TOTAL :            156       0     4538     5126     9664       0
=================================================================================================
                                  DIVISION TOTAL :       156       0     4538     5126     9664       0
=================================================================================================
```

Figure A.1

Appendix A

```
PROJECT   : SAMPLE HOTEL PROJECT              LOCATION : KINGSTON, MASSACHUSETTS
ARCH/ENGR : HOWARD, FINE & HOWARD, INC.       OWNER    : XYZ DEVELOPMENT
QUANTITIES BY: PLM                            ENTERED BY: PLM
```

3 CONCRETE

3.1 CONCRETE FORMWORK

DESCRIPTION LINE NO.	CREW	QUANTITY	UNIT	M/H	MATERIAL	LABOR	EQUIP	TOTAL	SUB
COLUMN FORMWORK, PLYWOOD, 16" @SQU. 4 USES									
031 142 6150 00 M	C1	7384.00	S.F.C.A	0.136	0.58	4.05	0.10	4.73	
-		100% 100%			1005.48	4304.87	29915.57	705.09	34925.54	
FORMS IN PLACE, FLOOR SLAB 19" METAL DOMES, 4 USES									
031 150 4150 00 M	C2	27720.00	S.F.	0.097	1.00	2.99	0.06	4.05	
-		100% 100%			2688.00	27747.72	82853.06	1675.52	112278.29	
FORMS IN PLACE, CONTINUOUS WALL FOOTINGS 4 USES									
031 158 0150 00 M	C1	1962.00	S.F.C.A	0.066	0.31	1.96	0.05	2.32	
-		100% 100%			129.45	604.30	3851.51	90.78	4546.58	
FORMS IN PLACE, SPREAD FOOTING, 4 USES									
031 158 5150 00 M	C1	1578.00	S.F.C.A	0.077	0.38	2.30	0.05	2.74	
-		100% 100%			121.97	607.53	3628.94	85.53	4322.00	
FORMS IN PLACE, JOB BUILT PLYWOOD, WALLS, TO 16' 4USE									
031 182 2550 00 M	C2	11172.00	S.F.C.A	0.122	0.66	3.75	0.08	4.48	
-		100% 100%			1357.61	7373.52	41846.03	846.24	50065.79	

```
                                 SUB TOTAL :    5303    40638   162095    3403   206136       0
```

3.2 CONCRETE REINFORCEMENT

DESCRIPTION LINE NO.	CREW	QUANTITY	UNIT	M/H	MATERIAL	LABOR	EQUIP	TOTAL	SUB
REINFORCING IN PLACE, COLUMNS, #8 TO #14									
032 107 0250 00 M	RODM4	20.50	TON	13.913	577.50	500.45	0.00	1077.95	
-		100% 100%			285.22	11838.75	10259.27	0.00	22098.02	
REINFORCING IN PLACE, ELEVATED SLABS, #4 TO #7									
032 107 0400 00 M	RODM4	40.20	TON	11.034	594.00	396.91	0.00	990.91	
-		100% 100%			443.59	23878.80	15955.80	0.00	39834.60	
REINFORCING IN PLACE FOOTINGS, #8 TO #14									
032 107 0550 00 M	RODM4	0.80	TON	8.889	566.50	319.73	0.00	886.23	
-		100% 100%			7.11	453.20	255.79	0.00	708.99	

Figure A.2

```
=================================================================================
NO. 1988-100            BURDENED ITEMIZED JOB REPORT                    PAGE  3
---------------------------------------------------------------------------------

PROJECT   : SAMPLE HOTEL PROJECT           LOCATION : KINGSTON, MASSACHUSETTS
ARCH/ENGR : HOWARD, FINE & HOWARD, INC.    OWNER    : XYZ DEVELOPMENT
QUANTITIES BY: PLM                         ENTERED BY: PLM

=================================================================================
REINFORCING IN PLACE FOOTINGS, #8 TO #14
  032 107 0550 01  M          RODM4  1.00  TON    8.889   566.50   319.73    0.00    886.23
  -                                                8.89    566.50   319.73    0.00    886.23

REINFORCING IN PLACE WALLS, #3 TO #7
  032 107 0700 00  M          RODM4  2.60  TON   10.667   588.50   383.68    0.00    972.18
  -                                               27.73  1530.10   997.57    0.00   2527.67

UNLOAD & SORT REBAR, INCL CRANE
 *032 172 0001 00  C          C5    65.10  TON    0.767     0.00    26.51    6.62     33.13
  -                                               49.96     0.00  1725.83  430.85   2156.68


                              SUB TOTAL :          823    38267    29514     431    68212      0
---------------------------------------------------------------------------------
  3.3 CAST-IN-PLACE CONCRETE
---------------------------------------------------------------------------------
DESCRIPTION                   CREW  QUANTITY UNIT   M/H   MATERIAL   LABOR    EQUIP   TOTAL    SUB
LINE NO.

---------------------------------------------------------------------------------

PLACING CONCRETE INCL. VIBRATING, 18" COLUMNS, PUMPED
 *033 172 0600 00  M          C20   91.00  C.Y.   1.067    58.69    28.12    9.00     95.80
  -                                               97.07  5340.34  2558.91  818.82   8718.08

PLACE AND VIBRATE CONCRETE, ELEVATED SLAB, OVER 10" THICK, PUMPED
 *033 172 1600 00  M          C20  955.00  C.Y.   0.427    58.69    11.25    3.60     73.53
  -                                              407.47 56044.18 10741.80 3437.24  70223.21

PLACE AND VIBRATE CONCRETE, CONTINUOUS SHALLOW FOOTINGS, DIRECT CHUTE
 *033 172 1900 00  M          C6    71.00  C.Y.   0.400    58.69    10.32    0.47     69.48
  -                                               28.40  4166.64   732.76   33.58   4932.98

PLACE AND VIBRATE CONCRETE, SPREAD FOOTINGS UNDER 1@C@Y, DIRECT CHUTE
 *033 172 2400 00  M          C6    34.00  C.Y.   0.873    58.69    22.52    1.03     82.23
  -                                               29.67  1995.29   765.60   35.04   2795.98

PLACING CONCRETE, INCL. VIBRATING, WALLS 8" THICK PUMPED
 *033 172 4950 00  M          C20  132.00  C.Y.   0.753    58.69    19.85    6.35     84.89
  -                                               99.39  7746.42  2620.11  838.40  11204.93

FINISHING FLOOR MONOLT DARBY FINISH
  033 454 0050 00  M          CEFI1 26724.00 S.F.  0.011    0.00     0.31    0.00      0.31
  -                                              285.06     0.00  8263.77    0.00   8263.77

                              SUB TOTAL :          947    75293    25683    5163   106139      0
=================================================================================
                              DIVISION TOTAL :    7072   154198   217292    8997   380487      0
=================================================================================
                              JOB TOTAL :         7228   154198   221830   14124   390151      0
=================================================================================
```

Figure A.3

```
===============================================================================================
NO. 1988-100              BURDENED DIVISION SUMMARY                              PAGE  1
-----------------------------------------------------------------------------------------------

PROJECT    : SAMPLE HOTEL PROJECT            LOCATION  : KINGSTON, MASSACHUSETTS
ARCH/ENGR  : HOWARD, FINE & HOWARD, INC.     OWNER     : XYZ DEVELOPMENT
QUANTITIES BY: PLM                           ENTERED BY: PLM

===============================================================================================

                             2 SITEWORK

-----------------------------------------------------------------------------------------------
SECTION NO.    DESCRIPTION                     MANHOURS  MATERIAL   LABOR    EQUIP    TOTAL    SUB
                                                                                     COST  CONTRACT
===============================================================================================

2.2            EARTHWORK                          156        0      4538     5126     9664      0

-----------------------------------------------------------------------------------------------

                              DIVISION TOTAL :    156        0      4538     5126     9664      0
===============================================================================================

                             3 CONCRETE

-----------------------------------------------------------------------------------------------
SECTION NO.    DESCRIPTION                     MANHOURS  MATERIAL   LABOR    EQUIP    TOTAL    SUB
                                                                                     COST  CONTRACT
===============================================================================================

3.1            CONCRETE FORMWORK                 5303    40637    162096     3403   206137      0

3.2            CONCRETE REINFORCEMENT             823    38268     29514      431    68213      0

3.3            CAST-IN-PLACE CONCRETE             947    75293     25683     5163   106139      0

-----------------------------------------------------------------------------------------------

                              DIVISION TOTAL :   7072   154198    217293     8998   380489      0
-----------------------------------------------------------------------------------------------
                              JOB TOTAL :        7228   154198    221831    14124   390153      0
===============================================================================================
```

Figure A.4

Logic Diagrams   Figures B.1 through B.6 are the logic diagrams for the sample hotel project. They represent the construction sequences for the various phases of the work. Note that each phase is contained on a separate sheet for clarity. The amount of record keeping is reduced by using a "standard" logic sequence for repetitive situations such as the tower sequence.

Schedule Reports   Figures B.7 through B.25 are computer print-outs from the sample hotel project. They represent the kind of information that can be derived from a well organized and coded schedule. Each is explained below.

Summary Bar Chart of the Entire Project
Figure B.7 represents a summary bar chart of all activities in the project obtained using the *SUMM* code that was designed for this purpose. This chart is especially useful for representing the overall plan to company management.

Schedule Reports by Project Phase
Figures B.8 through B.21 are schedule reports by project phase. Each of the major phases of the building is shown using a classic schedule report (also known as a *planning schedule* in some systems), and by a daily interval bar chart. Each report was printed by selecting activities with *AREA* codes equal to the appropriate phase of construction, and all were sorted by *ES-EF*. The foundation phase was summarized by major structural element, as shown in Figure B.10. These examples represent what is probably the most commonly used type of working schedule.

Other Types of Reports
Other possibilities include the bar chart of the electrical subcontractor's work (Figure B.22), which could be used for overall planning discussions with major subcontractors early in the job. Figure B.23 is a *TS-ES* sort of the foundation work activities, useful for separating critical activities from those which have some flexibility. Finally, Figures B.24 and B.25 are examples of phase schedules reflecting the progress that might have been achieved during the first month of work on the foundation. These schedules show actual work versus planned work.

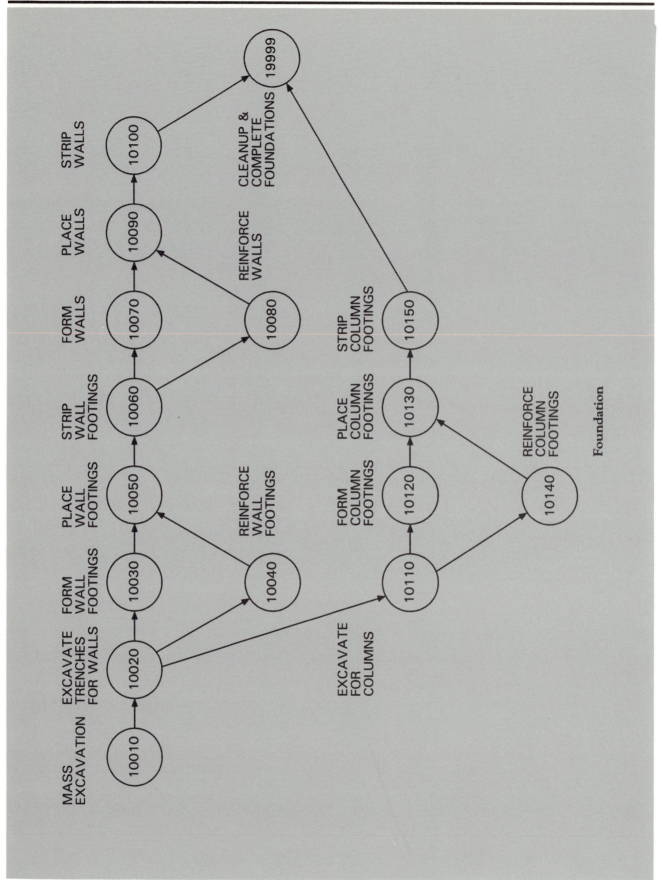

Figure B.1

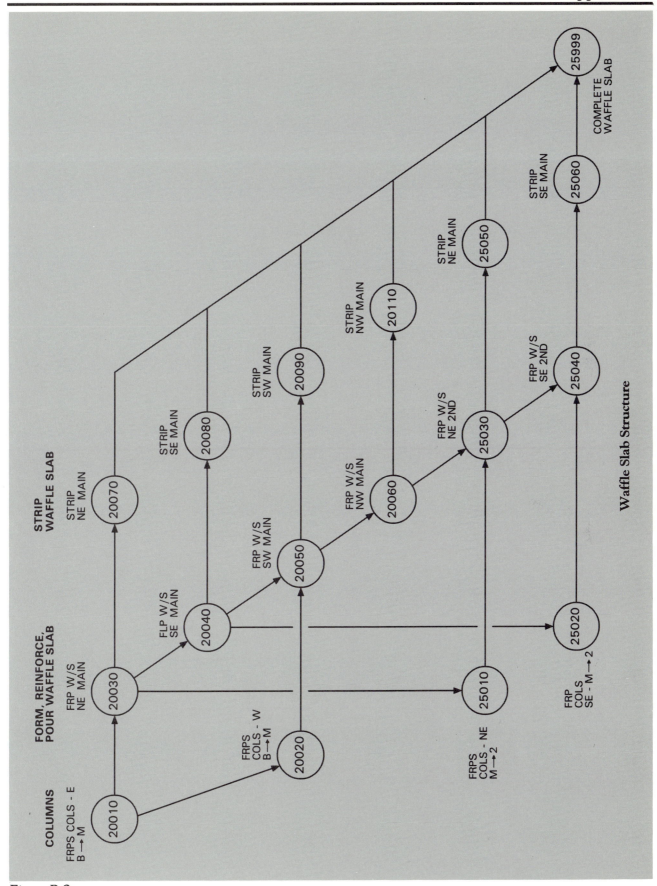

Waffle Slab Structure

Figure B.2

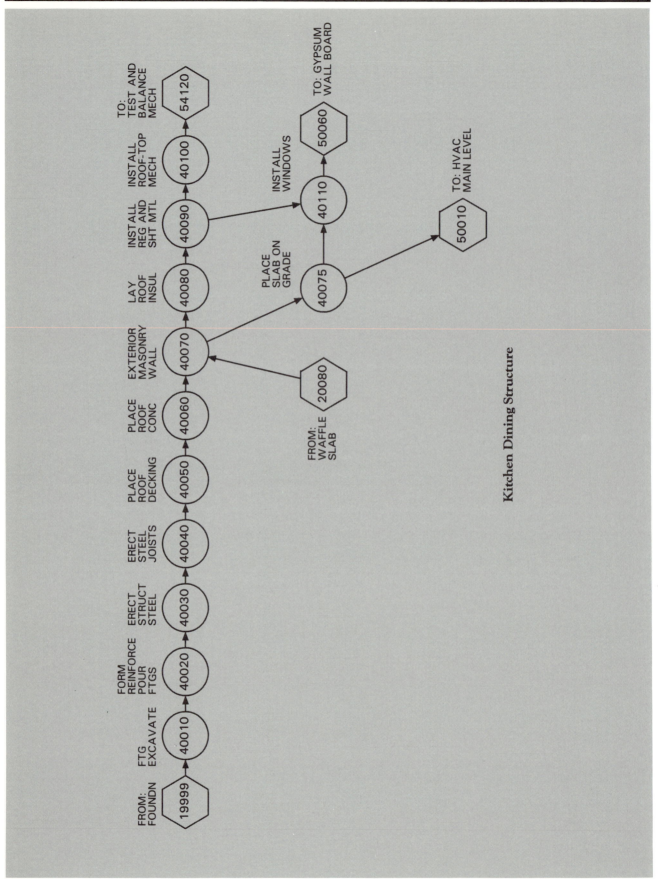

Kitchen Dining Structure

Figure B.3

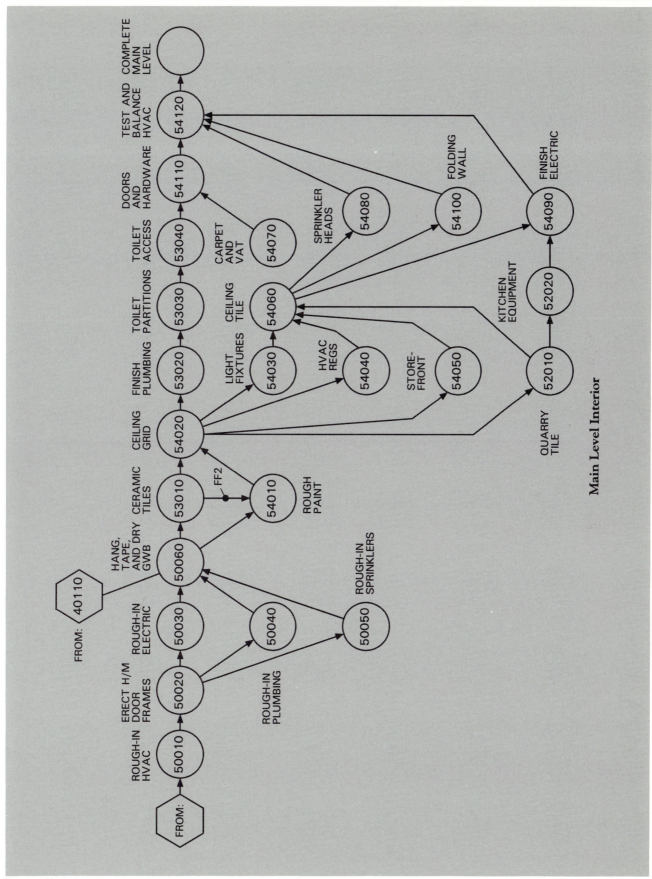

Main Level Interior

Figure B.4

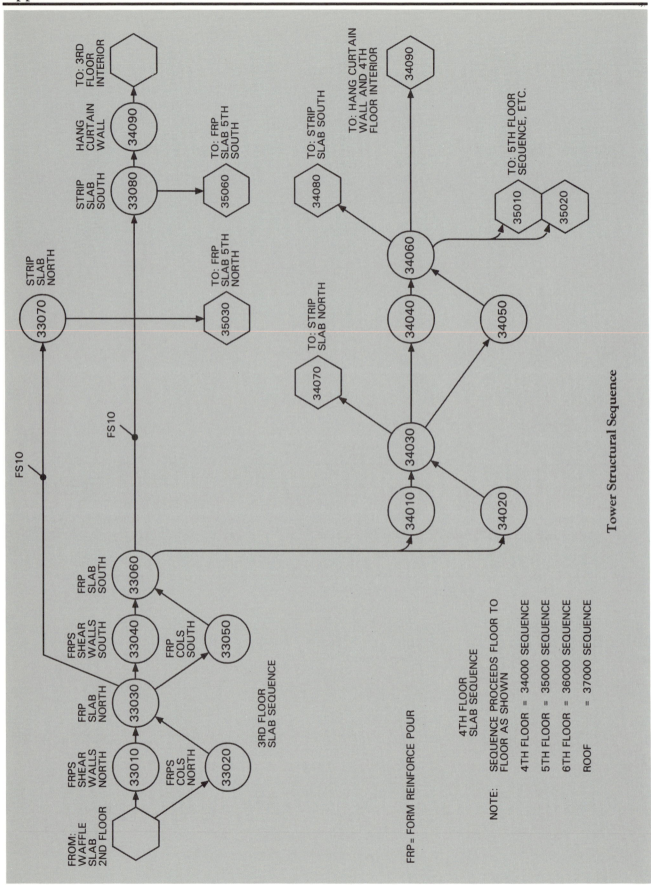

Tower Structural Sequence

Figure B.5

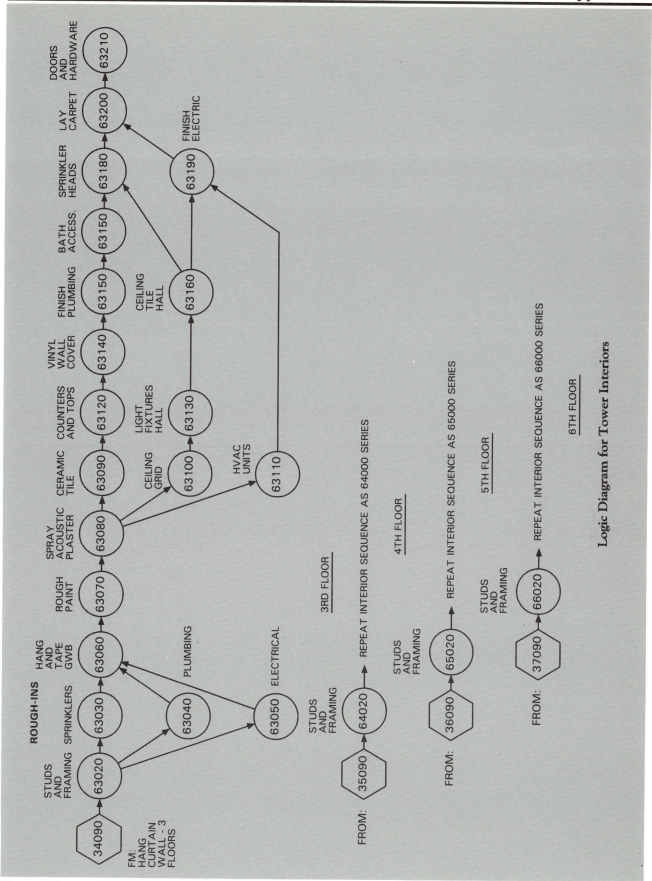

Logic Diagram for Tower Interiors

Figure B.6

```
-------------------------------------------------------------------------------------------------------------------------------
                                                        PRIMAVERA PROJECT PLANNER                       SAMPLE HOTEL PROJECT

REPORT DATE 18FEB88  RUN NO.   10                     SAMPLE HOTEL PROJECT SCHEDULE                    START DATE  1FEB88  FIN DATE 10APR89

SUMMARY BARCHART BY PHASE                                                                              DATA DATE  1FEB88   PAGE NO.    1

                                                                                                                   WEEKLY-TIME PER.   1
-------------------------------------------------------------------------------------------------------------------------------
 .............ACTIVITY DESCRIPTION.............                          07  04  02   06  04  01   05  03   07  05  02   06  06  03  01   05
ACTIVITY ID  OD   RD  PCT    CODES      FLOAT    SCHEDULE              MAR APR MAY  JUN JUL AUG  SEP OCT  NOV DEC JAN  FEB MAR APR MAY  JUN
-----------  ---- ---- --- ------------ -----    --------               88  88  88   88  88  88   88  88   88  88  89   89  89  89  89   89
                                                                    ---------------------------------------------------------------------
 FOUNDATION                                       CURRENT          EEEEEEEEEEEEEE    .   .   .    .   .    .   .   .    .   .   .   .    .

 WAFFLE SLAB STRUCTURE                            CURRENT          *    .  EEEEEEEEEEEEE .   .    .   .    .   .   .    .   .   .   .    .

 TOWER STRUCTURE                                  CURRENT          *    .   .   .    . EEEEEEEEEEEEEEEEEEEEEEEEEEEEEEEEEEE  .   .   .    .

 DINING AREA STRUCTURE                            CURRENT          *    .  EEEEEEEEEEEEEEEEE .    .   .    .   .   .    .   .   .   .    .

 MAIN LEVEL INTERIOR                              CURRENT          *    .   .   .    .   .EEEEEEEEEEEEEEEE .   .   .    .   .   .   .    .

 TOWER INTERIORS                                  CURRENT          *    .   .   .    .   .   .    . EEEEEEEEEEEEEEEEEEEEEEEEEEEEEEE  .    .
```

Figure B.7

PRIMAVERA PROJECT PLANNER   SAMPLE HOTEL PROJECT

REPORT DATE 18FEB88 RUN NO. 2   SAMPLE HOTEL PROJECT SCHEDULE   START DATE 1FEB88   FIN DATE 10APR89

CLASSIC SCHEDULE REPORT - SORTED BY ES, TF   DATA DATE 1FEB88   PAGE NO. 1

ACTIVITY ID	ORIG DUR	REM DUR	PCT	CODE	ACTIVITY DESCRIPTION	EARLY START	EARLY FINISH	LATE START	LATE FINISH	TOTAL FLOAT
10010	7	7	0	EXCVBSMT	MASS EXCAVATION	1FEB88	9FEB88	1FEB88	9FEB88	0
10020	2	2	0	EXCVBSMT	EXCAV TRENCH/WALL FOOTINGS	10FEB88	11FEB88	10FEB88	11FEB88	0
10110	1	1	0	EXCVBSMT	EXCAVATE COLUMN FOOTINGS	12FEB88	12FEB88	22MAR88	22MAR88	27
10030	6	6	0	GNCNBSMT03	FORM WALL FOOTINGS	12FEB88	19FEB88	12FEB88	19FEB88	0
10040	1	1	0	GNCNBSMT03	REINFORCE WALL FOOTINGS	15FEB88	19FEB88	12FEB88	19FEB88	0
10120	4	4	0	GNCNBSMT03	REINFORCE COLUMN FOOTINGS	15FEB88	18FEB88	23MAR88	28MAR88	30
10130	1	1	0	GNCNBSMT01	FORM COLUMN FOOTINGS	19FEB88	19FEB88	29MAR88	29MAR88	27
10050	1	1	0	GNCNBSMT01	PLACE WALL FOOTINGS	22FEB88	22FEB88	22FEB88	22FEB88	0
10150	2	2	0	GNCNBSMT03	PLACE COLUMN FOOTINGS	22FEB88	23FEB88	30MAR88	31MAR88	27
10060	3	3	0	GNCNBSMT03	STRIP FORMS COLUMN FOOTINGS	23FEB88	25FEB88	23FEB88	25FEB88	0
10080	3	3	0	GNCNBSMT03	STRIP WALL FOOTINGS	26FEB88	1MAR88	10MAR88	14MAR88	9
10070	12	12	0	GNCNBSMT03	REINFORCE BASEMENT WALLS	26FEB88	14MAR88	26FEB88	14MAR88	0
10090	3	3	0	GNCNBSMT01	FORM BASEMENT WALLS	15MAR88	17MAR88	15MAR88	17MAR88	0
10100	6	6	0	GNCNBSMT03	PLACE BASEMENT WALLS	18MAR88	25MAR88	18MAR88	25MAR88	0
10160	4	4	0	EXCVBSMT	BACKFILL WALLS BASEMENT	28MAR88	31MAR88	28MAR88	31MAR88	0
19999	0	0	0	BSMT	COMPLETE FOUNDATION	1APR88	1APR88	1APR88	1APR88	0

Figure B.8

```
                        PRIMAVERA PROJECT PLANNER                        SAMPLE HOTEL PROJECT

REPORT DATE 18FEB88  RUN NO.  16     SAMPLE HOTEL PROJECT SCHEDULE        START DATE  1FEB88   FIN DATE 10APR89

DAILY BARCHART SCHEDULE                                                  DATA DATE   1FEB88   PAGE NO.     1

                                                                         DAILY-TIME PER.   1
```

.........ACTIVITY DESCRIPTION....... ACTIVITY ID OD RD PCT CODES	FLOAT	SCHEDULE	01 FEB 88	08 FEB 88	15 FEB 88	22 FEB 88	29 FEB 88	07 MAR 88	14 MAR 88	21 MAR 88	28 MAR 88	04 APR 88	11 APR 88	18 APR 88	25 APR 88	02 MAY 88	09 MAY 88
MASS EXCAVATION 10010 7	0	EARLY	EEEEEEE														
EXCAV TRENCH/WALL FOOTINGS 10020 2	0	EARLY	*	EE													
EXCAVATE COLUMN FOOTINGS 10110 1	27	EARLY	*	.	E.												
FORM WALL FOOTINGS 10030 6	0	EARLY	*	.	EEEEEE.												
REINFORCE WALL FOOTINGS 10040 6	0	EARLY	*	.	EEEEEE.												
REINFORCE COLUMN FOOTINGS 10140 1	30	EARLY	*	.	E												
FORM COLUMN FOOTINGS 10120 4	27	EARLY	*	.	EEEE												
PLACE COLUMN FOOTINGS 10130 1	27	EARLY	*	.	.	E.											
PLACE WALL FOOTINGS 10050 1	0	EARLY	*	.	.	E											
STRIP FORMS COLUMN FOOTINGS 10150 2	27	EARLY	*	.	.	EE											
STRIP WALL FOOTINGS 10060 3	0	EARLY	*	.	.	.EEE											
REINFORCE BASEMENT WALLS 10080 3	9	EARLY	*	.	.	.	EEE										
FORM BASEMENT WALLS 10070 12	0	EARLY	*	.	.	.	EEEEEEEEEEE										
PLACE BASEMENT WALLS 10090 3	0	EARLY	*	.	.	.	.	.	.EEE								
STRIP FORMS BASEMENT WALLS 10100 3	0	EARLY	*	.	.	.	.	.	.	EEEEEE.							
BACKFILL WALLS BASEMENT 10160 6	0	EARLY	*	.	.	.	.	.	.	.	EEEE						
COMPLETE FOUNDATION 19999 4	0	EARLY	*	.	.	.	.	.	.	.	.	E.					
COMPLETE FOUNDATION 19999 0	0	EARLY	*	.	.	.	.	.	.	.	.	E.					

Figure B.9

```
                    PRIMAVERA PROJECT PLANNER              SAMPLE HOTEL PROJECT

REPORT DATE 18FEB88 RUN NO.  33    SAMPLE HOTEL PROJECT SCHEDULE       START DATE  1FEB88  FIN DATE 10APR89

DAILY BARCHART SCHEDULE                                               DATA DATE   1FEB88  PAGE NO.    1

                                                                              DAILY-TIME PER.   1

.........ACTIVITY DESCRIPTION.........                01  08  15  22  29  07  14  21  28  04  11  18  25  02  09
ACTIVITY ID  OD  RD  PCT   CODES   FLOAT   SCHEDULE    FEB FEB FEB FEB FEB MAR MAR MAR MAR APR APR APR APR MAY MAY
-----------  --  --  ---   -----   -----   --------    88  88  88  88  88  88  88  88  88  88  88  88  88  88  88

C06                                         CURRENT    *   .   .   .   .   .   .   .  EEEEE.  .   .   .   .   .
                                                       **  .   .   .   .   .   .   .   .   .   .   .   .   .   .

FOUNDATION EXCAVATION                       CURRENT    EEEEEEEEE.  .   .   .   .   .   .   .   .   .   .   .   .
                                                       *   .   .   .   .   .   .   .   .   .   .   .   .   .   .

WALL FOOTINGS                               CURRENT    *   .   EEEEEEEEE.  .   .   .   .   .   .   .   .   .   .
                                                       **  .   .   .   .   .   .   .   .   .   .   .   .   .   .

COLUMN FOOTINGS                             CURRENT    *   .   EEEEEE  .   .   .   .   .   .   .   .   .   .   .
                                                       **  .   .   .   .   .   .   .   .   .   .   .   .   .   .

FOUNDATION WALLS                            CURRENT    *   .   .   .   EEEEEEEEEEEEEEEEEEEE.  .   .   .   .   .
                                                       **  .   .   .   .   .   .   .   .   .   .   .   .   .   .
```

Figure B.10

PRIMAVERA PROJECT PLANNER

SAMPLE HOTEL PROJECT SCHEDULE

SAMPLE HOTEL PROJECT

REPORT DATE 18FEB88   RUN NO.   17

DAILY BARCHART SCHEDULE

START DATE 1FEB88   FIN DATE 10APR89

DATA DATE 1FEB88   PAGE NO.   1

DAILY-TIME PER.   1

ACTIVITY ID	OD	RD	PCT	CODES	FLOAT	SCHEDULE	...ACTIVITY DESCRIPTION...

Date scale across top: 04 APR 88, 11 APR 88, 18 APR 88, 25 APR 88, 02 MAY 88, 09 MAY 88, 16 MAY 88, 23 MAY 88, 30 MAY 88, 06 JUN 88, 13 JUN 88, 20 JUN 88, 27 JUN 88, 04 JUL 88

Activity Description	Activity ID	Float	Schedule	Bar position
FRPS W/S COLS,EAST, BSMT TO MAIN	20010	0	EARLY	EEE (04 APR)
FRPS W/S COLS-WEST, BSMT TO MAIN	20020	13	EARLY	EEE. (11 APR)
FRP WFFL SLBS-NE, MAIN LVL	20030	0	EARLY	EEEEEEE. (04–11 APR)
FRPS COLS-NE, MAIN TO 2FLR	25010	17	EARLY	EEE (18 APR)
FRP WFFLE SLBS-SE, MAIN LEVEL	20040	0	EARLY	EEEEEEEE (18–25 APR)
FRPS COLS-SE, MAIN TO 2FLR	25020	17	EARLY	EEE (02 MAY)
FRP WFFLE SLBS-SW, MAIN LVL	20050	0	EARLY	EEEEEE (02–09 MAY)
FRP WFFLE SLBS-NW, MAIN LVL	20060	0	EARLY	EEEEEEE. (09–16 MAY)
FRP WFFLE SLBS-NE, 2FLR	25030	0	EARLY	EEEEEEEE (16–23 MAY)
STRP WFFLE SLBS-SW, MAIN	20090	22	EARLY	EE (23 MAY)
FRP WFFLE SLB-SE, 2FLR	25040	0	EARLY	EEEEEEEE (30 MAY–06 JUN)
STRP WFFLE SLBS-NW, MAIN	20100	16	EARLY	EE (06 JUN)
STRP WFFLE SLB-NE, MAIN LVL	20070	8	EARLY	EE. (13 JUN)
STRP WFFLE SLB-SE, MAIN LVL	20080	0	EARLY	EE (20 JUN)
STRP WFFLE SLB-SE, MAIN LVL	25050	0	EARLY	EE (20 JUN)
STRP WFFLE SLB-SE, 2FLR	25060	0	EARLY	EE (20 JUN)
COMPLETE WAFFLE SLAB	25999	0	EARLY	E (27 JUN)

Figure B.11

252

```
                                         PRIMAVERA PROJECT PLANNER          SAMPLE HOTEL PROJECT

REPORT DATE 18FEB88  RUN NO.    4              SAMPLE HOTEL PROJECT SCHEDULE        START DATE 1FEB88  FIN DATE 10APR89

CLASSIC SCHEDULE REPORT - SORTED BY ES, TF                                         DATA DATE  1FEB88    PAGE NO.   1
```

ACTIVITY ID	ORIG DUR	REM DUR	PCT	CODE	ACTIVITY DESCRIPTION	EARLY START	EARLY FINISH	LATE START	LATE FINISH	TOTAL FLOAT
20010	3	3	0	GNCNBSMT03	FRPS W/S COLS,EAST, BSMT TO MAIN	1APR88	5APR88	1APR88	5APR88	0
20020	3	3	0	GNCNBSMT03	FRPS W/S COLS-WEST, BSMT TO MAIN	6APR88	8APR88	25APR88	27APR88	13
20030	8	8	0	GNCNMAIN03	FRP WFFL SLBS-NE, MAIN LVL	6APR88	15APR88	6APR88	15APR88	0
25010	3	3	0	GNCNMAIN03	FRPS COLS-NE, MAIN TO 2FLR	18APR88	20APR88	11MAY88	13MAY88	17
20040	8	8	0	GNCNMAIN03	FRP WFFLE SLBS-SE, MAIN LEVEL	18APR88	27APR88	18APR88	27APR88	0
25020	3	3	0	GNCNMAIN03	FRPS COLS-SE, MAIN TO 2FLR	28APR88	2MAY88	23MAY88	25MAY88	17
20050	6	6	0	GNCNMAIN03	FRP WFFLE SLBS-SW, MAIN LVL	28APR88	5MAY88	28APR88	5MAY88	0
20060	6	6	0	GNCNMAIN03	FRP WFFLE SLBS-NW, MAIN LVL	6MAY88	13MAY88	6MAY88	13MAY88	0
25030	8	8	0	GNCN2FLR03	FRP WFFLE SLBS-NE, 2FLR	16MAY88	25MAY88	16MAY88	25MAY88	0
20090	2	2	0	GNCNMAIN03	STRP WFFLE SLBS-SW, MAIN	20MAY88	23MAY88	21JUN88	22JUN88	22
25040	8	8	0	GNCN2FLR03	FRP WFFLE SLB-SE, 2FLR	26MAY88	6JUN88	26MAY88	6JUN88	0
20100	2	2	0	GNCNMAIN03	STRP WFFLE SLBS-NW, MAIN	30MAY88	31MAY88	21JUN88	22JUN88	16
20070	2	2	0	GNCNMAIN03	STRP WFFLE SLB-NE, MAIN LVL	9JUN88	10JUN88	21JUN88	22JUN88	8
20080	2	2	0	GNCNMAIN03	STRP WFFLE SLB-SE, MAIN LVL	21JUN88	22JUN88	21JUN88	22JUN88	0
25050	2	2	0	GNCNMAIN03	STRP WFFLE SLB-SE, MAIN LVL	21JUN88	22JUN88	21JUN88	22JUN88	0
25060	2	2	0	GNCN2FLR03	STRP WFFLE SLB-SE, 2FLR	21JUN88	22JUN88	21JUN88	22JUN88	0
25999	0	0	0	GNCN	COMPLETE WAFFLE SLAB	23JUN88	23JUN88	23JUN88	23JUN88	0

Figure B.12

```
                              PRIMAVERA PROJECT PLANNER          SAMPLE HOTEL PROJECT

REPORT DATE 18FEB88  RUN NO.   2        SAMPLE HOTEL PROJECT SCHEDULE     START DATE  1FEB88  FIN DATE 10APR89

CLASSIC SCHEDULE REPORT - SORTED BY ES, TF                                DATA DATE  1FEB88  PAGE NO.   2
```

ACTIVITY ID	ORIG DUR	REM DUR	PCT	CODE	ACTIVITY DESCRIPTION	EARLY START	EARLY FINISH	LATE START	LATE FINISH	TOTAL FLOAT
40010	1	1	0	EXCVMAIN	EXCAV FOOTING - K/D AREA	1APR88	1APR88	2NOV88	2NOV88	153
40020	9	9	0	GNCNMAINO3	FRPS FOOTINGS - K/D AREA	4APR88	14APR88	3NOV88	15NOV88	153
40030	2	2	0	MAIN	ERECT STRUC STEEL - K/D AREA	15APR88	18APR88	16NOV88	17NOV88	153
40040	4	4	0	MAIN	ERECT STEEL JOISTS K/D AREA	19APR88	22APR88	18NOV88	23NOV88	153
40050	5	5	0	MAIN	PLACE STEEL ROOF DECKING K/D AREA	25APR88	29APR88	24NOV88	30NOV88	153
40060	1	1	0	GNCNMAINO3	PLACE ROOF CONCRETE K/D AREA	2MAY88	2MAY88	1DEC88	1DEC88	153
40070	10	10	0	MAIN	ERECT EXT MASONRY WALL - MAIN LVL	23JUN88	6JUL88	2DEC88	15DEC88	116
40080	2	2	0	MAIN	LAY ROOF INSULATION K/D AREA	7JUL88	8JUL88	27MAR89	28MAR89	187
40075	5	5	0	GNCNMAINO3	PLACE SOG - K/D AREA	7JUL88	13JUL88	16DEC88	22DEC88	116
40090	4	4	0	MAIN	INSTALL ROOFING & SHT MTL K/D AREA	11JUL88	14JUL88	29MAR89	3APR89	187
40110	7	7	0	MAIN	INSTALL WINDOWS MAIN LEVEL	14JUL88	22JUL88	25JAN89	2FEB89	139
40100	3	3	0	HVACMAIN	INSTALL ROOF TOP MECHANICAL K/D AREA	15JUL88	19JUL88	4APR89	6APR89	187

Figure B.13

```
                    PRIMAVERA PROJECT PLANNER                    SAMPLE HOTEL PROJECT

REPORT DATE 18FEB88  RUN NO.  19      SAMPLE HOTEL PROJECT SCHEDULE        START DATE 1FEB88  FIN DATE 10APR89

DAILY BARCHART SCHEDULE                                                   DATA DATE 1FEB88   PAGE NO.   1

                                                                                      DAILY-TIME PER.   1

......ACTIVITY DESCRIPTION......                          04   11   18   25   02   09   16   23   30   06   13   20   27   04
ACTIVITY ID  OD RD PCT  CODES   FLOAT   SCHEDULE          APR  APR  APR  APR  MAY  MAY  MAY  MAY  MAY  JUN  JUN  JUN  JUN  JUL
                                                          88   88   88   88   88   88   88   88   88   88   88   88   88   88

EXCAV FOOTING - K/D AREA
   40010   1                      153   CURRENT           E.
FRPS FOOTINGS - K/D AREA
   40020   9                      153   CURRENT           EEEEEEEEE
ERECT STRUC STEEL - K/D AREA
   40030   2                      153   CURRENT              EE
ERECT STEEL JOISTS K/D AREA
   40040   4                      153   CURRENT               EEEE
PLACE STEEL ROOF DECKING K/D AREA
   40050   5                      153   CURRENT                   EEEEE
PLACE ROOF CONCRETE K/D AREA
   40060   1                      153   CURRENT                         E
ERECT EXT MASONRY WALL - MAIN LVL
   40070  10                      116   CURRENT                                                 EEEEEEEEEE
LAY ROOF INSULATION K/D AREA
   40080   2                      187   CURRENT                                                             EE
PLACE SOG - K/D AREA
   40075   5                      116   CURRENT
INSTALL ROOFING & SHT MTL K/D AREA
   40090   4                      187   CURRENT                                                             EE
INSTALL WINDOWS MAIN LEVEL
   40110   7                      139   CURRENT
INSTALL ROOF TOP MECHANICAL K/D AREA
   40100   3                      187   CURRENT
```

Figure B.14

255

PRIMAVERA PROJECT PLANNER

SAMPLE HOTEL PROJECT

REPORT DATE 18FEB88 RUN NO. 19

START DATE 1FEB88 FIN DATE 10APR89

DAILY BARCHART SCHEDULE

SAMPLE HOTEL PROJECT SCHEDULE

DATA DATE 1FEB88 PAGE NO. 1

DAILY-TIME PER. 2

...........ACTIVITY DESCRIPTION......							11 JUL 88	18 JUL 88	25 JUL 88	01 AUG 88	08 AUG 88	15 AUG 88	22 AUG 88	29 AUG 88	05 SEP 88	12 SEP 88	19 SEP 88	26 SEP 88	03 OCT 88	10 OCT 88	17 OCT 88
ACTIVITY ID	OD	RD	PCT	CODES	FLOAT	SCHEDULE															
EXCAV FOOTING - K/D AREA 40010 1					153	CURRENT	.	.	.	.	.	.	.	.	.	.	.	.	.	.	.
FRPS FOOTINGS - K/D AREA 40020 9					153	CURRENT	.	.	.	.	.	.	.	.	.	.	.	.	.	.	.
ERECT STRUC STEEL - K/D AREA 40030 2					153	CURRENT	.	.	.	.	.	.	.	.	.	.	.	.	.	.	.
ERECT STEEL JOISTS K/D AREA 40040 4					153	CURRENT	.	.	.	.	.	.	.	.	.	.	.	.	.	.	.
PLACE STEEL ROOF DECKING K/D AREA 40050 5					153	CURRENT	.	.	.	.	.	.	.	.	.	.	.	.	.	.	.
PLACE ROOF CONCRETE K/D AREA 40060 1					153	CURRENT	.	.	.	.	.	.	.	.	.	.	.	.	.	.	.
ERECT EXT MASONRY WALL - MAIN LVL 40070 10					116	CURRENT	.	.	.	.	.	.	.	.	.	.	.	.	.	.	.
LAY ROOF INSULATION K/D AREA 40080 2					187	CURRENT	.	.	.	.	.	.	.	.	.	.	.	.	.	.	.
PLACE SOG - K/D AREA 40075 5					116	CURRENT	EEE	.	.	.	.	.	.	.	.	.	.	.	.	.	.
INSTALL ROOFING & SHT MTL K/D AREA 40090 4					187	CURRENT	EEEE	.	.	.	.	.	.	.	.	.	.	.	.	.	.
INSTALL WINDOWS MAIN LEVEL 40110 7					139	CURRENT	.	EEEEEEE	.	.	.	.	.	.	.	.	.	.	.	.	.
INSTALL ROOF TOP MECHANICAL K/D AREA 40100 3					187	CURRENT	.	EEE	.	.	.	.	.	.	.	.	.	.	.	.	.

Figure B.15

PRIMAVERA PROJECT PLANNER

SAMPLE HOTEL PROJECT

REPORT DATE 18FEB88 RUN NO. 5   SAMPLE HOTEL PROJECT SCHEDULE   START DATE 1FEB88 FIN DATE 10APR89

CLASSIC SCHEDULE REPORT - SORTED BY ES, TF   DATA DATE 1FEB88   PAGE NO. 1

ACTIVITY ID	ORIG DUR	REM DUR	PCT	CODE	ACTIVITY DESCRIPTION	EARLY START	EARLY FINISH	LATE START	LATE FINISH	TOTAL FLOAT
33020	3	3	0	GNCN3FLR03	FRP COLS-N	23JUN88	27JUN88	28JUN88	30JUN88	3
33010	6	6	0	GNCN3FLR03	FRPS SHR WALLS-N	23JUN88	30JUN88	23JUN88	30JUN88	0
33030	7	7	0	GNCN3FLR03	FRP SLAB-N	1JUL88	11JUL88	1JUL88	11JUL88	0
33050	3	3	0	GNCN3FLR03	FRPS COLS-S	12JUL88	14JUL88	15JUL88	19JUL88	3
33040	6	6	0	GNCN3FLR03	FRPS SHR WALLS-S	12JUL88	19JUL88	12JUL88	19JUL88	0
33060	7	7	0	GNCN3FLR03	FRP SLAB-S	20JUL88	28JUL88	20JUL88	28JUL88	0
33070	3	3	0	GNCN3FLR03	STRP SLAB-N	26JUL88	28JUL88	5JAN89	9JAN89	117
33080	3	3	0	GNCN3FLR03	STRP SLAB-S	5SEP88	7SEP88	5JAN89	9JAN89	88
34090	7	7	0	3FLR	HANG CURTAIN WALL - 3FLR	14OCT88	24OCT88	10JAN89	18JAN89	62

Figure B.16

257

```
                    PRIMAVERA PROJECT PLANNER                        SAMPLE HOTEL PROJECT

REPORT DATE 18FEB88  RUN NO.  26      SAMPLE HOTEL PROJECT SCHEDULE     START DATE 1FEB88   FIN DATE 10APR89

DAILY BARCHART SCHEDULE                                                DATA DATE  1FEB88   PAGE NO.   1

                                                                       DAILY-TIME PER.     1
```

........ACTIVITY DESCRIPTION.......					06 JUN 88	13 JUN 88	20 JUN 88	27 JUN 88	04 JUL 88	11 JUL 88	18 JUL 88	25 JUL 88	01 AUG 88	08 AUG 88	15 AUG 88	22 AUG 88	29 AUG 88	05 SEP 88
ACTIVITY ID	OD RD PCT CODES	SCHEDULE	FLOAT															
FRP COLS-N 33020	3	CURRENT	3		.	.	.	EEE	.	.	.	.	.	.	.	.	.	.
FRPS SHR WALLS-N 33010	6	CURRENT	0		.	.	.	EEEEEE	.	.	.	.	.	.	.	.	.	.
FRP SLAB-N 33030	7	CURRENT	0		.	.	.	.	EEEEEEE	.	.	.	.	.	.	.	.	.
FRPS COLS-S 33050	3	CURRENT	3		.	.	.	.	.	EEE	.	.	.	.	.	.	.	.
FRPS SHR WALLS-S 33040	6	CURRENT	0		.	.	.	.	.	EEEEE	EEEEE	.	.	.	.	.	.	.
FRP SLAB-S 33060	7	CURRENT	0		.	.	.	.	.	.	EEEEE	EEEEEEE	.	.	.	.	.	.
STRP SLAB-N 33070	3	CURRENT	117		.	.	.	.	.	.	.	.EEE	.	.	.	.	.	.
STRP SLAB-S 33080	3	CURRENT	88		.	.	.	.	.	.	.	.	.	.	.	.	.	.
HANG CURTAIN WALL - 3FLR 34090	7	CURRENT	62		.	.	.	.	.	.	.	.	.	.	.	EEE	.	.

Figure B.17

PRIMAVERA PROJECT PLANNER

SAMPLE HOTEL PROJECT

REPORT DATE 18FEB88 RUN NO. 2   SAMPLE HOTEL PROJECT SCHEDULE   START DATE 1FEB88 FIN DATE 10APR89

CLASSIC SCHEDULE REPORT - SORTED BY ES, TF   DATA DATE 1FEB88   PAGE NO. 3

ACTIVITY ID	ORIG DUR	REM DUR	PCT	CODE	ACTIVITY DESCRIPTION	EARLY START	EARLY FINISH	LATE START	LATE FINISH	TOTAL FLOAT
50010	10	10	0	HVACMAIN	ROUGH IN HVAC-MAIN LVL	14JUL88	27JUL88	23DEC88	5JAN89	116
50020	10	10	0	STUDMAIN	ERECT HM FRMES/STUDS - MAIN LVL	28JUL88	10AUG88	6JAN89	19JAN89	116
50050	6	6	0	MAIN	ROUGH-IN SPNKLR PIPING, MAIN LVL	11AUG88	18AUG88	26JAN89	2FEB89	120
50030	8	8	0	ELECMAIN	ROUGH IN ELEC - MAIN LVL	11AUG88	22AUG88	24JAN89	2FEB89	118
50040	10	10	0	PLMBMAIN	ROUGH PLUMBING - MAIN LVL	11AUG88	24AUG88	20JAN89	2FEB89	116
50060	10	10	0	GWBSMAIN	HANG & TAPE GWB - MAIN LVL	25AUG88	7SEP88	3FEB89	16FEB89	116
53010	3	3	0	MAIN	INSTALL CERAMIC TILE - MAIN LVL	8SEP88	12SEP88	24FEB89	28FEB89	121
54010	8	8	0	PANTMAIN	ROUGH PAINT & VWC - MAIN LVL	8SEP88	19SEP88	17FEB89	28FEB89	116
54020	7	7	0	STUDMAIN	INSTALL CEILING GRID - MAIN LVL	20SEP88	28SEP88	1MAR89	9MAR89	116
54050	3	3	0	MAIN	INSTALL STORE FRONT - MAIN LVL	29SEP88	30SEP88	21MAR89	23MAR89	123
52010	5	5	0	MAIN	LAY QUARRY TILE - MAIN LEVEL	29SEP88	5OCT88	10MAR89	16MAR89	116
54040	5	5	0	HVACMAIN	FINISH HVAC - MAIN LVL	29SEP88	5OCT88	17MAR89	23MAR89	121
54030	7	7	0	ELECMAIN	INSTALL LIGHT FIXTURES - MAIN LVL	29SEP88	7OCT88	15MAR89	23MAR89	119
53020	8	8	0	PLMBMAIN	FINISH PLUMBING - MAIN LVL	29SEP88	10OCT88	17MAR89	28MAR89	121
52020	10	10	0	GNCNMAIN06	INSTALL KITCHEN EQUIP. - MAIN LVL	6OCT88	19OCT88	17MAR89	30MAR89	116
54060	3	3	0	GWBSMAIN	INSTALL CEILING TILES - MAIN LVL	10OCT88	12OCT88	24MAR89	28MAR89	119
53030	2	2	0	GNCNMAIN06	INSTALL TOILET PARTITIONS -MAIN LVL	11OCT88	12OCT88	29MAR89	30MAR89	121
53040	2	2	0	GNCNMAIN06	INSTALL BATH ACCESS.- MAIN LVL	13OCT88	14OCT88	31MAR89	3APR89	121
54100	2	2	0	GNCNMAIN06	INSTALL FOLDING WALL - MAIN LVL	13OCT88	14OCT88	5APR89	6APR89	124
54080	3	3	0	MAIN	INSTALL SPRINKLER HEADS - MAIN LVL	13OCT88	17OCT88	4APR89	6APR89	123
54070	4	4	0	CRPTMAIN	LAY CARPET-MAIN LVL	13OCT88	18OCT88	29MAR89	3APR89	119
54110	3	3	0	GNCNMAIN06	HANG DOORS - MAIN LVL	19OCT88	21OCT88	4APR89	6APR89	119
54090	5	5	0	ELECMAIN	FINISH ELECTRIC - MAIN LVL	20OCT88	26OCT88	31MAR89	6APR89	118
54120	2	2	0	HVACMAIN	TEST & BALANCE HVAC MAIN LVL	27OCT88	28OCT88	7APR89	10APR89	116

Figure B.18

PRIMAVERA PROJECT PLANNER

SAMPLE HOTEL PROJECT

REPORT DATE 18FEB88   RUN NO. 30   SAMPLE HOTEL PROJECT SCHEDULE

START DATE 1FEB88   FIN DATE 10APR89

DAILY BARCHART SCHEDULE   DATA DATE 1FEB88   PAGE NO. 1

DAILY-TIME PER. 1

| ....ACTIVITY DESCRIPTION........ | | | | CODES | FLOAT | SCHEDULE | 04 JUL 88 | 11 JUL 88 | 18 JUL 88 | 25 JUL 88 | 01 AUG 88 | 08 AUG 88 | 15 AUG 88 | 22 AUG 88 | 29 AUG 88 | 05 SEP 88 | 12 SEP 88 | 19 SEP 88 | 26 SEP 88 | 03 OCT 88 |
ACTIVITY ID OD RD PCT																				
ROUGH IN HVAC-MAIN LVL 50010 10				HVAC	116	CURRENT		EEEEEEEEE												
ERECT HM FRMES/STUDS - MAIN LVL 50020 10				STUD	116	CURRENT					EEEEEEEEE									
ROUGH-IN SPNKLR PIPING, MAIN LVL 50050 6					120	CURRENT							EEEEE							
ROUGH IN ELEC - MAIN LVL 50030 8				ELEC	118	CURRENT							EEEEEEE							
ROUGH PLUMBING - MAIN LVL 50040 10				PLMB	116	CURRENT							EEEEEEEEE							
HANG & TAPE GWB - MAIN LVL 50060 10				GWBS	116	CURRENT									EEEEEEEEE					
INSTALL CERAMIC TILE - MAIN LVL 53010 3					121	CURRENT											EEE			
ROUGH PAINT & VWC - MAIN LVL 54010 8				PANT	116	CURRENT								EEEEEEEE						
INSTALL CEILING GRID - MAIN LVL 54020 7				STUD	116	CURRENT										EEEEEEE				
INSTALL STORE FRONT - MAIN LVL 54050 3					123	CURRENT												EEE		
LAY QUARRY TILE - MAIN LEVEL 52010 5					116	CURRENT													EEEEE	
FINISH HVAC - MAIN LVL 54040 5				HVAC	121	CURRENT													EEEEE	
INSTALL LIGHT FIXTURES - MAIN LVL 54030 7				ELEC	119	CURRENT													EEEEEEE	
FINISH PLUMBING - MAIN LVL 53020 8				PLMB	121	CURRENT													EEEEEEE	

Figure B.19

REPORT DATE 18FEB88 RUN NO. 9   PRIMAVERA PROJECT PLANNER   SAMPLE HOTEL PROJECT

SAMPLE HOTEL PROJECT SCHEDULE   START DATE 1FEB88 FIN DATE 10APR89

CLASSIC SCHEDULE REPORT - SORTED BY ES, TF   DATA DATE 1FEB88 PAGE NO. 1

ACTIVITY ID	ORIG DUR	REM DUR	PCT	CODE	ACTIVITY DESCRIPTION	EARLY START	EARLY FINISH	LATE START	LATE FINISH	TOTAL FLOAT
63020	8	8	0	STUD3FLR	INSTALL DOOR FRAMES & STUDS	25OCT88	3NOV88	19JAN89	30JAN89	62
63030	5	5	0	3FLR	ROUGH-IN SPRNKLRS	4NOV88	10NOV88	2FEB89	8FEB89	64
63040	5	5	0	PLMB3FLR	ROUGH PLUMBING	4NOV88	10NOV88	2FEB89	8FEB89	64
63050	7	7	0	ELEC3FLR	ROUGH IN ELECTRIC	4NOV88	14NOV88	31JAN89	8FEB89	62
63060	7	7	0	GWBS3FLR	HANG & TAPE GWB	15NOV88	23NOV88	9FEB89	17FEB89	62
63070	2	2	0	PANT3FLR	ROUGH PAINT	24NOV88	25NOV88	20FEB89	21FEB89	62
63080	2	2	0	GWBS3FLR	SPRAY ACOUS PLSTR	28NOV88	29NOV88	22FEB89	23FEB89	62
63100	2	2	0	STUD3FLR	INSTALL CEILING GRID	30NOV88	1DEC88	23MAR89	24MAR89	81
63090	3	3	0	3FLR	INSTALL CER. TILE	30NOV88	2DEC88	24FEB89	28FEB89	62
63110	3	3	0	HVAC3FLR	INSTALL HVAC UNITS	30NOV88	2DEC88	24MAR89	28MAR89	82
63130	1	1	0	ELEC3FLR	INSTALL LIGHT FIXTURES	2DEC88	2DEC88	27MAR89	27MAR89	81
63160	1	1	0	GWBS3FLR	INSTALL CEILING TILES	5DEC88	5DEC88	28MAR89	28MAR89	81
63120	4	4	0	GNCN3FLR06	INSTALL COUNTER TOPS	5DEC88	8DEC88	1MAR89	6MAR89	62
63190	2	2	0	ELEC3FLR	FINISH ELECTRIC	6DEC88	7DEC88	29MAR89	30MAR89	81
63140	6	6	0	3FLR	INSTALL VWC	9DEC88	16DEC88	7MAR89	14MAR89	62
63150	5	5	0	PLMB3FLR	FINISH PLUMBING	19DEC88	23DEC88	15MAR89	21MAR89	62
63170	5	5	0	GNCN3FLR06	INSTALL BATHROOM ACCESSORIES	26DEC88	30DEC88	22MAR89	28MAR89	62
63180	2	2	0	3FLR	INSTALL SPRNKLR HEADS	2JAN89	3JAN89	29MAR89	30MAR89	62
63200	4	4	0	CRPT3FLR	LAY CARPET	4JAN89	9JAN89	31MAR89	5APR89	62
63210	3	3	0	GNCN3FLR06	HANG DOORS & HARDWARE	10JAN89	12JAN89	6APR89	10APR89	62

Figure B.20

Figure B.21

```
                                    PRIMAVERA PROJECT PLANNER                    SAMPLE HOTEL PROJECT

REPORT DATE 18FEB88  RUN NO.  22    SAMPLE HOTEL PROJECT SCHEDULE                START DATE 1FEB88  FIN DATE 10APR89

DAILY BARCHART SCHEDULE                                                          DATA DATE 1FEB88   PAGE NO.  1

                                                                                      DAILY-TIME PER.   1
```

...ACTIVITY DESCRIPTION...						03 OCT 88	10 OCT 88	17 OCT 88	24 OCT 88	31 OCT 88	07 NOV 88	14 NOV 88	21 NOV 88	28 NOV 88	05 DEC 88	12 DEC 88	19 DEC 88	26 DEC 88	02 JAN 89	09 JAN 89
ACTIVITY ID	OD	RD	PCT	CODES	SCHEDULE FLOAT															
INSTALL DOOR FRAMES & STUDS 63020	8				CURRENT 62				. EEEEEEEE											
ROUGH-IN SPRNKLRS 63030	5				CURRENT 64						EEEEE									
ROUGH PLUMBING 63040	5				CURRENT 64						EEEEE									
ROUGH IN ELECTRIC 63050	7				CURRENT 62						EEEEEEE									
HANG & TAPE GWB 63060	7				CURRENT 62							EEEEEEE								
ROUGH PAINT 63070	2				CURRENT 62								EE.							
SPRAY ACOUS PLSTR 63080	2				CURRENT 62									EE						
INSTALL CEILING GRID 63100	2				CURRENT 81									EE						
INSTALL CER. TILE 63090	3				CURRENT 62									EEE.						
INSTALL HVAC UNITS 63110	3				CURRENT 82									EEE.						
INSTALL LIGHT FIXTURES 63130	1				CURRENT 81									E.						
INSTALL CEILING TILES 63160	1				CURRENT 81									E						
INSTALL COUNTER TOPS 63120	4				CURRENT 62									EEEE						
FINISH ELECTRIC 63190	2				CURRENT 81									.EE						
INSTALL VWC 63140	6				CURRENT 62										EEEEEE.					
FINISH PLUMBING 63150	5				CURRENT 62											EEEEE.				
INSTALL BATHROOM ACCESSORIES 63170	5				CURRENT 62												EEEEE.			
INSTALL SPRNKLR HEADS 63180	2				CURRENT 62												.	EE	
LAY CARPET 63200	4				CURRENT 62															
HANG DOORS & HARDWARE 63210	3				CURRENT 62															:E
```
262
```

PRIMAVERA PROJECT PLANNER   SAMPLE HOTEL PROJECT

REPORT DATE 18FEB88 RUN NO. 29   SAMPLE HOTEL PROJECT SCHEDULE

DAILY BARCHART SCHEDULE   START DATE 1FEB88   FIN DATE 10APR89

DATA DATE 1FEB88   PAGE NO. 1

WEEKLY-TIME PER. 1

......ACTIVITY DESCRIPTION......							01 AUG 88	05 SEP 88	03 OCT 88	07 NOV 88	05 DEC 88	02 JAN 89	06 FEB 89	06 MAR 89	03 APR 89	01 MAY 89	05 JUN 89	03 JUL 89	07 AUG 89	04 SEP 89	02 OCT 89	06 NOV 89	04 DEC 89
ACTIVITY ID	OD	RD	PCT	CODES	FLOAT	SCHEDULE																	
ROUGH IN ELECTRIC 62050	7			2FLR	88	CURRENT		EE															
INSTALL LIGHT FIXTURES 62130	1			2FLR	107	CURRENT			E.														
FINISH ELECTRIC 62190	2			2FLR	107	CURRENT			E.														
ROUGH IN ELECTRIC 63050	7			3FLR	62	CURRENT			EEE														
INSTALL LIGHT FIXTURES 63130	1			3FLR	81	CURRENT				E.													
FINISH ELECTRIC 63190	2			3FLR	81	CURRENT				E													
ROUGH IN ELECTRIC 64050	7			4FLR	36	CURRENT				EE													
INSTALL LIGHT FIXTURES 64130	1			4FLR	55	CURRENT					E												
FINISH ELECTRIC 64190	2			4FLR	55	CURRENT					EE												
ROUGH IN ELECTRIC 65050	7			5FLR	10	CURRENT						EE											
INSTALL LIGHT FIXTURES 65130	1			5FLR	29	CURRENT							E										
FINISH ELECTRIC 65190	2			5FLR	29	CURRENT							E										
ROUGH IN ELECTRIC 66050	7			6FLR	0	CURRENT							EE										
INSTALL LIGHT FIXTURES 66130	1			6FLR	19	CURRENT								E.									
FINISH ELECTRIC 66190	2			6FLR	19	CURRENT								E.									
ROUGH-IN ELECTRIC BASEMENT 29020	5			BSMT	245	CURRENT	EEE																
ROUGH IN ELEC - MAIN LVL 50030	8			MAIN	118	CURRENT		EE															
INSTALL LIGHT FIXTURES - MAIN LVL 54030	7			MAIN	119	CURRENT			EE														
FINISH ELECTRIC - MAIN LVL 54090	5			MAIN	116	CURRENT			EE														

Figure B.22

PRIMAVERA PROJECT PLANNER   SAMPLE HOTEL PROJECT

REPORT DATE 18FEB88 RUN NO. 24   SAMPLE HOTEL PROJECT SCHEDULE   START DATE 1FEB88 FIN DATE 10APR89
DAILY BARCHART SCHEDULE   DATA DATE 1FEB88   PAGE NO. 1
                                                                        DAILY-TIME-PER. 1

..........ACTIVITY DESCRIPTION.......... ACTIVITY ID	OD	RD	PCT	CODES	FLOAT	SCHEDULE	01 FEB 88	08 FEB 88	15 FEB 88	22 FEB 88	29 FEB 88	07 MAR 88	14 MAR 88	21 MAR 88	28 MAR 88	04 APR 88	11 APR 88	18 APR 88	25 APR 88	02 MAY 88	09 MAY 88
MASS EXCAVATION 10010	7				0	CURRENT	/EEEEE														
EXCAV TRENCH/WALL FOOTINGS 10020	2				0	CURRENT	**	/E													
FORM WALL FOOTINGS 10030	6				0	CURRENT	**	/EEEEE													
REINFORCE WALL FOOTINGS 10040	6				0	CURRENT	**	/EEEEE													
PLACE WALL FOOTINGS 10050	1				0	CURRENT	**	/													
STRIP WALL FOOTINGS 10060	3				0	CURRENT	**		./EE												
FORM BASEMENT WALLS 10070	12				0	CURRENT	**			/EE											
PLACE BASEMENT WALLS 10090	3				0	CURRENT	**				/EEEEEEEEEE										
STRIP FORMS BASEMENT WALLS 10100	6				0	CURRENT	**						.EE							
BACKFILL WALLS BASEMENT 10160	4				0	CURRENT	**							/EEEEE.						
COMPLETE FOUNDATION 19999	0				0	CURRENT	**								/EEE					
REINFORCE BASEMENT WALLS 10080	3				9	CURRENT	**				EEE+++++LLL			/						
EXCAVATE COLUMN FOOTINGS 10110	1				27	CURRENT	**	E++++++++++++++++++++++L													
FORM COLUMN FOOTINGS 10120	4				27	CURRENT	**	EEEE+++++++++++++++++LLLL													
PLACE COLUMN FOOTINGS 10130	1				27	CURRENT	**	E+++++++++++++++++++++++L													
STRIP FORMS COLUMN FOOTINGS 10150	2				27	CURRENT	**	EE++++++++++++++++++++++LL													
REINFORCE COLUMN FOOTINGS 10140	1				30	CURRENT	**	E++++++++++++++++++++++++L													

Figure B.23

```
                              PRIMAVERA PROJECT PLANNER          SAMPLE HOTEL PROJECT

REPORT DATE 18FEB88 RUN NO.  2      SAMPLE HOTEL PROJECT SCHEDULE       START DATE  1FEB88  FIN DATE  7APR89

SCHEDULE REPORT COMPARISON TO TARGET SCHEDULE                   DATA DATE 29FEB88    PAGE NO.   1

ACTIVITY  TAR  CUR  CUR                                      CURRENT     EARLY     TARGET     EARLY
   ID     DUR  DUR  PCT   CODE   ACTIVITY DESCRIPTION        START       FINISH    START      FINISH    VAR.

 10010     7    5   100         MASS EXCAVATION              3FEB88A     9FEB88A    1FEB88     9FEB88     0
 10020     1    5   100         EXCAV TRENCH/WALL FOOTINGS   8FEB88A    12FEB88A   10FEB88    10FEB88    -2
 10030     4    6   100         FORM WALL FOOTINGS          10FEB88A    17FEB88A   11FEB88    16FEB88    -1
 10110     1    2   100         EXCAVATE COLUMN FOOTINGS    12FEB88A    14FEB88A   11FEB88    11FEB88    -2
 10040     1    3   100         REINFORCE WALL FOOTINGS     13FEB88A    17FEB88A   11FEB88    11FEB88    -4
 10120     3    4   100         FORM COLUMN FOOTINGS        17FEB88A    20FEB88A   12FEB88    16FEB88    -4
 10140     1    2   100         REINFORCE COLUMN FOOTINGS   19FEB88A    20FEB88A   12FEB88    12FEB88    -6
 10050     1    1   100         PLACE WALL FOOTINGS         22FEB88A    23FEB88A   17FEB88    17FEB88    -4
 10150     2    2   100         STRIP FORMS COLUMN FOOTINGS 24FEB88A    25FEB88A   18FEB88    19FEB88    -4
 10060     2    1    67         STRIP WALL FOOTINGS         26FEB88A    29FEB88    18FEB88    19FEB88    -6
 10070    11   10    17         FORM BASEMENT WALLS         27FEB88A    11MAR88    22FEB88     7MAR88    -4
 10130     1    0   100         PLACE COLUMN FOOTINGS                   22FEB88A   17FEB88    17FEB88    -3
 10080     2    3     0         REINFORCE BASEMENT WALLS     1MAR88      3MAR88    22FEB88    23FEB88    -7
 10090     2    3     0         PLACE BASEMENT WALLS        14MAR88     16MAR88     8MAR88     9MAR88    -5
 10100    11    6     0         STRIP FORMS BASEMENT WALLS  17MAR88     24MAR88    10MAR88    24MAR88     0
 10160     4    4     0         BACKFILL WALLS BASEMENT     25MAR88     30MAR88    25MAR88    30MAR88     0
 19999     0    0     0         COMPLETE FOUNDATION         31MAR88     31MAR88    31MAR88    31MAR88     0
```

Figure B.24

PRIMAVERA PROJECT PLANNER   SAMPLE HOTEL PROJECT

REPORT DATE 18FEB88 RUN NO. 1   SAMPLE HOTEL PROJECT SCHEDULE   START DATE 1FEB88 FIN DATE 7APR89

BAR CHART - CURRENT VS. TARGET COMPARISON   DATA DATE 29FEB88   PAGE NO. 1

DAILY-TIME PER. 1

......ACTIVITY DESCRIPTION......						SCHEDULE	01 FEB 88	08 FEB 88	15 FEB 88	22 FEB 88	29 FEB 88	07 MAR 88	14 MAR 88	21 MAR 88	28 MAR 88	04 APR 88	11 APR 88	18 APR 88	25 APR 88	02 MAY 88	09 MAY 88
ACTIVITY ID	OD	RD	PCT	CODES	FLOAT																

MASS EXCAVATION 10010 7 0 100 — CURRENT / TARGET

EXCAV TRENCH/WALL FOOTINGS 10020 2 0 100 — CURRENT / TARGET

FORM WALL FOOTINGS 10030 6 0 100 — CURRENT / TARGET

EXCAVATE COLUMN FOOTINGS 10110 1 0 100 — CURRENT / TARGET

REINFORCE WALL FOOTINGS 10040 6 0 100 — CURRENT / TARGET

FORM COLUMN FOOTINGS 10120 4 0 100 — CURRENT / TARGET

REINFORCE COLUMN FOOTINGS 10140 1 0 100 — CURRENT / TARGET

PLACE WALL FOOTINGS 10050 1 0 100 — CURRENT / TARGET

STRIP FORMS COLUMN FOOTINGS 10150 2 0 100 — CURRENT / TARGET

STRIP WALL FOOTINGS 10060 3 1 67 — CURRENT / TARGET

FORM BASEMENT WALLS 10070 12 10 17 — CURRENT / TARGET

PLACE COLUMN FOOTINGS 10130 1 0 100 — CURRENT / TARGET

REINFORCE BASEMENT WALLS 10080 3 3 0 — CURRENT / TARGET

PLACE BASEMENT WALLS 10090 3 3 0 — CURRENT / TARGET

STRIP FORMS BASEMENT WALLS 10100 6 6 0 — CURRENT / TARGET

BACKFILL WALLS BASEMENT 10160 4 4 0 — CURRENT / TARGET

COMPLETE FOUNDATION 19999 0 0 0 — CURRENT / TARGET

Figure B.25

C. S. I. MASTERFORMAT BROADSCOPE HEADINGS

Division 1 — General Requirements

01010	Summary of Work
01020	Allowances
01025	Measurement and Payment
01030	Alternates/Alternatives
01040	Coordination
01050	Field Engineering
01060	Regulatory Requirements
01070	Abbreviations and Symbols
01080	Identification Systems
01090	Reference Standards
01100	Special Project Procedures
01200	Project Meetings
01300	Submittals
01400	Quality Control
01500	Construction Facilities and Temporary Controls
01600	Material and Equipment
01650	Starting of Systems/Commissioning
01700	Contract Closeout
01800	Maintenance

Division 2 — Sitework

02010	Subsurface Investigation
02050	Demolition
02100	Site Preparation
02140	Dewatering
02150	Shoring and Underpinning
02160	Excavation Support Systems
02170	Cofferdams
02200	Earthwork
02300	Tunneling
02350	Piles and Caissons
02450	Railroad Work
02480	Marine Work
02500	Paving and Surfacing
02600	Piped Utility Materials
02660	Water Distribution
02680	Fuel Distribution
02700	Sewerage and Drainage
02760	Restoration of Underground Pipelines
02770	Ponds and Reservoirs
02780	Power and Communications
02800	Site Improvements
02900	Landscaping

Division 3 — Concrete

03100 Concrete Formwork
03200 Concrete Reinforcement
03250 Concrete Accessories
03300 Cast-in-Place Concrete
03370 Concrete Curing
03400 Precast Concrete
03500 Cementitious Decks
03600 Grout
03700 Concrete Restoration and Cleaning
03800 Mass Concrete

Division 4 — Masonry

04100 Mortar
04150 Masonry Accessories
04200 Unit Masonry
04400 Stone
04500 Masonry Restoration and Cleaning
04550 Refractories
04600 Corrosion Resistant Masonry

Division 5 — Metals

05010 Metal Materials
05030 Metal Finishes
05050 Metal Fastening
05100 Structural Metal Framing
05200 Metal Joists
05300 Metal Decking
05400 Cold-Formed Metal Framing
05500 Metal Fabrications
05580 Sheet Metal Fabrications
05700 Ornamental Metal
05800 Expansion Control
05900 Hydraulic Structures

Division 6 — Wood and Plastics

06050 Fasteners and Adhesives
06100 Rough Carpentry
06130 Heavy Timber Construction
06150 Wood-Metal Systems
06170 Prefabricated Structural Wood
06200 Finish Carpentry
06300 Wood Treatment
06400 Architectural Woodwork
06500 Prefabricated Structural Plastics
06600 Plastic Fabrications

Division 7 — Thermal and Moisture Protection

07100	Waterproofing
07150	Dampproofing
07190	Vapor and Air Retarders
07200	Insulation
07250	Fireproofing
07300	Shingles and Roofing Tiles
07400	Preformed Roofing and Cladding/Siding
07500	Membrane Roofing
07600	Flashing and Sheet Metal
07700	Roof Specialties and Accessories
07800	Skylights
07900	Joint Sealers

Division 8 — Doors and Windows

08100	Metal Doors and Frames
08200	Wood and Plastic Doors
08250	Door Opening Assemblies
08300	Special Doors
08400	Entrances and Storefronts
08500	Metal Windows
08600	Wood and Plastic Windows
08650	Special Windows
08700	Hardware
08800	Glazing
08900	Glazed Curtain Walls

Division 9 — Finishes

09100	Metal Support Systems
09200	Lath and Plaster
09230	Aggregate Coatings
09250	Gypsum Board
09300	Tile
09400	Terrazzo
09500	Acoustical Treatment
09540	Special Surfaces
09550	Wood Flooring
09600	Stone Flooring
09630	Unit Masonry Flooring
09650	Resilient Flooring
09680	Carpet
09700	Special Flooring
09780	Floor Treatment
09800	Special Coatings
09900	Painting
09950	Wall Coverings

Division 10 — Specialties

10100	Chalkboards and Tackboards
10150	Compartments and Cubicles
10200	Louvers and Vents
10240	Grilles and Screens
10250	Service Wall Systems
10260	Wall and Corner Guards
10270	Access Flooring
10280	Specialty Modules
10290	Pest Control
10300	Fireplaces and Stoves
10340	Prefabricated Exterior Specialties
10350	Flagpoles
10400	Identifying Devices
10450	Pedestrian Control Devices
10500	Lockers
10520	Fire Protection Specialties
10530	Protective Covers
10550	Postal Specialties
10600	Partitions
10650	Operable Partitions
10670	Storage Shelving
10700	Exterior Sun Control Devices
10750	Telephone Specialties
10800	Toilet and Bath Accessories
10880	Scales
10900	Wardrobe and Closet Specialties

Division 11 — Equipment

11010	Maintenance Equipment
11020	Security and Vault Equipment
11030	Teller and Service Equipment
11040	Ecclesiastical Equipment
11050	Library Equipment
11060	Theater and Stage Equipment
11070	Instrumental Equipment
11080	Registration Equipment
11090	Checkroom Equipment
11100	Mercantile Equipment
11110	Commercial Laundry and Dry Cleaning Equipment
11120	Vending Equipment
11130	Audio-Visual Equipment
11140	Service Station Equipment
11150	Parking Control Equipment
11160	Loading Dock Equipment
11170	Solid Waste Handling Equipment
11190	Detention Equipment
11200	Water Supply and Treatment Equipment
11280	Hydraulic Gates and Valves
11300	Fluid Waste Treatment and Disposal Equipment
11400	Food Service Equipment

11450   Residential Equipment
11460   Unit Kitchens
11470   Darkroom Equipment
11480   Athletic, Recreational and Therapeutic Equipment
11500   Industrial and Process Equipment
11600   Laboratory Equipment
11650   Planetarium Equipment
11660   Observatory Equipment
11700   Medical Equipment
11780   Mortuary Equipment
11850   Navigation Equipment

Division 12 — Furnishings

12050   Fabrics
12100   Artwork
12300   Manufactured Casework
12500   Window Treatment
12600   Furniture and Accessories
12670   Rugs and Mats
12700   Multiple Seating
12800   Interior Plants and Planters

Division 13 — Special Construction

13010   Air Supported Structures
13020   Integrated Assemblies
13030   Special Purpose Rooms
13080   Sound, Vibration, and Seismic Control
13090   Radiation Protection
13100   Nuclear Reactors
13120   Pre-Engineered Structures
13150   Pools
13160   Ice Rinks
13170   Kennels and Animal Shelters
13180   Site Constructed Incinerators
13200   Liquid and Gas Storage Tanks
13220   Filter Underdrains and Media
13230   Digestion Tank Covers and Appurtenances
13240   Oxygenation Systems
13260   Sludge Conditioning Systems
13300   Utility Control Systems
13400   Industrial and Process Control Systems
13500   Recording Instrumentation
13550   Transportation Control Instrumentation
13600   Solar Energy Systems
13700   Wind Energy Systems
13800   Building Automation Systems
13900   Fire Suppression and Supervisory Systems

Division 14 — Conveying Systems

14100   Dumbwaiters
14200   Elevators
14300   Moving Stairs and Walks
14400   Lifts
14500   Material Handling Systems
14600   Hoists and Cranes
14700   Turntables
14800   Scaffolding
14900   Transportation Systems

Division 15 — Mechanical

15050   Basic Mechanical Materials and Methods
15250   Mechanical Insulation
15300   Fire Protection
15400   Plumbing
15500   Heating, Ventilating, and Air Conditioning (HVAC)
15550   Heat Generation
15650   Refrigeration
15750   Heat Transfer
15850   Air Handling
15880   Air Distribution
15950   Controls
15990   Testing, Adjusting, and Balancing

Division 16 — Electrical

16050   Basic Electrical Materials and Methods
16200   Power Generation
16300   High Voltage Distribution (Above 600-Volt)
16400   Service and Distribution (600-Volt and Below)
16500   Lighting
16600   Special Systems
16700   Communications
16850   Electric Resistance Heating
16900   Controls
16950   Testing

Index

A

"Accelerating" the schedule, 195
Accounting systems, 10, 207
Activities
  adjusting times, 87-89
  administrative and support, 54, 57
  averaging, 89
  breaking the job down into, 53-57
  calculation of durations, 72-76
  code titles, 128
  codes dictionaries, 128,
  codes, samples, 128-130
  complex relationships, 61-63
  construction, monitoring. See Tracking
    system
  critical, status of, 154
  detail, level of, 55-56
  developing a list of, 56
  distribution of resources across,
    178-180
  estimating times, 166-167
  evaluating each independently, 70
  general types, 53-54
  information, 16, 48
  noncritical, 154-155
  overlapping, 62
  priority of relationships, 60-61
  procurement, 53-54, 57
  production, 53
  rounding, 89
  scheduling the procurement, 119-120
  shortening individual, 197-201
  sorting, 133-136
  specific types, 54-55
  submittal data, 111
  summary of, 136, 139
  time estimates, 69
  times, adjusted calculations, 76-87
  tips on list development, 57
  updating the individual, 145-147

Activity analysis records, 167
Activity Analysis Sheet, 71, 166, 170
Activity lists
  description of, 72
  examples of, 162-164
  making, 159
Administrative procedures for
  procurement. See Procurement
Agencies, public and government, 39
Arrow Diagramming Method (ADM), 57

B

Backward pass, 94-96
Bar charts
  displaying schedule with, 153
  for information for construction, 170
  product of the scheduling phase, 48
Benefits of using good management
  systems, 7-9
"Brainstorming", 57, 159
Budget calculations, 215, see also
  calculations
Building Construction Cost Data, 76
Building project, sample, introduction to,
  18
Buildings, vertically-oriented, 55
Buy-out, 37

C

Calculations. See also Job duration,
  advanced activity, 96-103
  budget, 215
  downstream impact, 146
  Expected Finish (ExF) formula, 147
  Percent Complete (PC) formula, 146
  Remaining Duration (RD) formula,
    146
  of required resource, 176-178
  trial, running the, 167-170

Calendar dates
  conversion of workdays to,   101-103
  establishing project,   101
  multiple approach,   103
  weather, dealing with the effects of,
    on,   103
Cash flow
  managing,   184-191
  forecasting,   188-191
Cash inflow curve,   191
Cash outflow curve,   188
Catalog cuts,   108
Coding schemes. *See also* Schedule
  by contractor or subcontractor,
    127-128
  other coding possibilities,   128
  for a project,   127-128
  by project level,   127
  by project phase,   127
  by trade,   127
  types of,   126
Communication
  and follow-up in procurement,   109
  in monitoring the project,   143-144
  in organizational changes,   10
  with project personnel,   31
  requirements for effective,   144
  as a task of manager,   6
Computer. *See also* Computer coding
  activities codes dictionary,   128
  applications,   10
  -based management systems,   8
  inputting data to the,   167
  processing data,   221
  software. *See* QWIKNET;
    PRIMAVERA
  use in the construction industry,
    8, 10-11
  value of, resources,   180, 182-183
  windows, programs that use,   167
Computer coding
  by digit position,   131-132
  groups,   128
  information,   128-132
  using the, structure to get information,
    132
Constrained dates
  definition of,   99
  early start/early finish (ES/EF),   99
  information about,   98-99
  no earlier than (NET),   99
  no later than (NLT),   99, 100
  start no earlier than (SNET),   99
Construction plan, developing the,   58
Construction projects
  aim of the book, ix,   195
  efficient management through sched-
    ules. *See* Schedule
Construction, providing information for,
  170-171
Construction schedule, coordinating with
  submittals,   111, *see also* Submittal
  data
Control, project. *See also* Planning,
  project

cycle, the,   15-28
  definition of,   4
  keeping good records,   42
  key to, by project management,   7
  monitor and,   143-156, *see also*
    Monitoring
  overall,   231
  as phase of scheduling,   48
  setting up for,   31
  as a task of manager,   6
  tasks,   231
Contract
  awarding,   159
  document review,   32
  keeping good records,   41
  owner requirements in,   38, *see also*
    Owners
  time allowed,   33
  "unwritten rules,"   38
Corrective action, taking,   155-156
Correspondence, keeping good records,
  41
Cost coding systems, project
  additional points about,   209
  detail in,   209
  elements of,   208-209
  information about,   208-209
Cost control, project
  See Equipment; Labor;
    Material; Subcontractors
  information about,   207-227
  issues,   225-227
  specific tasks in,   210-225
Costs. *See also* Cost control, project
  actual,   xv, 226
  budgeted,   xv
  depreciation,   226
  labor and equipment,   218
  management system element,   5
  overruns and cost, cause determination.
  as variable that affects project
    management,   9
CPM
  basic technique,   47
  calculation of duration,   96
  as a cost effective method,   231
  information about,   47-48
  pitfall of using,   49
"Crashing" the schedule,   195
Crew
  additional,   201
  composition,   34
  involvement,   54
Critical path,   196, *see also* CPM
Critical Path Method. *See* CPM
C.S.I. MASTERFORMAT,   71, 166,
  271-276
  "cushion" time,   81

D

Daily production rate method,   72
  calculations, basic,   76
  information about,   74-76

Data
  historical, collecting, 17
  inputting, to the computer, 167
  management of submittal, and
    procurement, 107-120, *see also*
    Procurement
  processing, 221-222
  providing, to the project team, 32
  scheduling to manage, 124
Data date, 145
Depreciation costs, 226
Designer's review, 109
Diagramming systems, 57
Decisions
  better decision-making, 9
  definition of, 4
Documentation, schedule, 170
Downstream impact, 145

E
Efficiency rate of production, 69
Equipment
  days or hours, 180
  estimator's assumptions about, 34
  recording and tracking costs, 218
Estimate
  the job, 210-212
  as source of information about subcon-
    tractor's work, 57
Estimate review, 33-39
Estimator meeting
  information, 33-39
  and the project team, 33-34
Expenses, covering, 188
Exception information, 154
Exception-oriented, as a management
  system, 6
Experience method of activities duration,
  76

F
Feedback cycle, 207
Feedback loop. *See also* Control, project
  definition of, 5
  points about, additional, 6
  and the project control cycle, 15
Finances, management system element,
  5
Floats
  definition, 96
  free, 100
  negative, 100
Formulas. *See* Calculations
Forward pass, 92

G
Geographic lines, 55
Goals
  comparing progress to, 152-153
  description of, 3
  of monitoring and controlling, 15
  setting, 4-5, 16

I
Inspection, and notification requirements,
  33

J
Job cost system, 221
Job delay, finding out about, 155
Job duration, calculating overall
  actual calculation procedure, 91-92
  backward pass, 91
  calculation of floats and critical path,
    92
  definitions of the calculation process,
    89-91
  demonstration of the actual procedure,
    92
  forward pass, 91
  goals of project calculation procedure,
    89
  information about, 89-96
Job estimating, 210-212
Job logic, 156, *see also* Arrow
  Diagramming Method (ADM);
  Precedence Diagramming Method
  (PDM)
Job logs, daily, 147-152
Job progress, determining, 152
Job site
  coordination on the, 40
  filing system, 41
Jobs, establishing plans in a three-step
  process, 16

L
Labor
  efficiency, promoting, 175
  paying, 184, *see also* Cash flow
  recording and tracking costs, 218
  as a resource, 175
  suppliers, 39
  time cards, 40
Laying off, 180
Logic diagram
  construction plan represented by, 57
  definition of, 16, 231
  result of planning phase, 48
  sample, 243-249

M
Management
  basic functions of, 16-17
  by exception, 8-9
  defined, 3-4
  information systems, 9
  procedures, setting up, 40
  resource. *See* Resource management
Management, project. *See* Project
  management
Manager, project. *See* Project manager
Managing the project, 5-6
Man-hour productivity method
  calculations, basic, 72-74
  information, 72-74
Manpower. *See* Crew

MASTERFORMAT, CSI, 71, 166, 271-276
Materials
   procurement and deliveries, 43
   as a resource, 175
Means Man-Hour Standards, 72
Monitoring. *See also* Control, project
   controlling and, 143-156
   effects of outside factors, 143
   as phase of scheduling, 48
   progress, 16-17, 143
   progress and schedule update,
      144-152, *see also* Schedule update
   steps in the process of, 143

N

Network
   construction plan represented by, 57
   expanding the, 58-65
   logic, changing the, 197
   result of planning phase, 48

O

Organizational concerns
   inflexibility, 10
   as variable that affects project
      management, 9
Owners, and their representatives, 38-39

P

Payment system. *See* Cash flow
Permits, 39
Personnel
   description of, 4
   identifying key, 32
   as variable that affects project
      management, 9
Plans, job. *See* Job plans
Planning, pre-construction. *See*
   Pre-construction planning
Planning, project. *See also* Control,
   project
   definition of, 4
   information about, 53-65
   monitoring the job, xv
   as a task of manager, 6
   a three-step process, xv
Planning schedules
   definition of, 243
   for information for construction, 170
   product of the planning phase, 48
   sample of, 115
Plugged dates, 152
Pre-construction planning, 31-43
Precedence Diagramming Method (PDM),
   57
Pre-planning process
   description of, 31
   with other parties, 36-39
   purpose of, 38
PRIMAVERA system software, 126
Problems, catching at an early stage, 6
Process
   definition of, 3
   information, 17

Procurement. *See also* Submittal data
   catalog cuts, 108
   key elements in successful, 108-110
   management of submittal data and,
      107-120
   problem, the source of the, 107-108
   procedures, basic, 108
   scheduling the, activities, 119-120
   shop drawing and, 107
   submittal data, 108
Productivity-based techniques and
   methods, 72
Productivity rate, basic, 85
Productivity tool, 48, *see also* CPM
Progress measurement, 143, *see also*
   Monitoring
Project administration, 36
Project captains, role of, 109
Project Control Cycle, 15, 231
Project control systems, 9
Project cost code, 208, *see also* Cost
   coding systems, project
"Project expediting," 195
Project management. *See also* Project
   manager
   approaches to, 5
   benefits, 7
   creating a system, 5
   develop a plan, 31
   elements of system, 5
   proactive, 5
   reactive, 5
   team attitude, 108
   variables that affect, 9-10
   what is, 3-11
Project manager. *See also* Project
   management
   in leadership role, 31
   responsibility, key, 233
   tasks of the, xv, 6, 31
Project, scheduling
   estimating durations, 69-87
   information about, 69-103
Project specifications, 110
Project team
   initial meeting, 35-36
   meeting with the estimators, 33-34
Project time
   guidelines for reducing, 201-203
   how to reduce, 196-201
   reasons for reducing overall, 195
   relationship of project cost to, 196
Projects, high-rise or multi-level, 127

Q

Quantity takeoffs, estimating, 35
QWIKNET scheduling system software,
   126

R

Rational process, in making decisions, 4
Record keeping. *See also* Tracking system
   procedures for setting up and
      maintaining, 110
   processing of documents, 111

setting up a good system,   41-43, 110
and tracking,   110-119
Records, schedule. *See* Schedule records
Regulatory agencies,   7
Relationships
  complex,   61-63
  finish-to-finish (FF),   96
  finish-to-start (FS),   61
  lag,   62
  priority of,   60-61
  start-to-start (SS),   62, 96
Reports
  keeping good records,   42, *see also*
    Tracking system
  project, what to look for,   154
  update,   171
Resource management
  and cash flow,   188-191, *see also* Cash
    flow
  information about,   175-192
  practical aspects of,   191-192
  process, development of,   175-176
Resources
  computer, value of,   180, 182-183
  calculations of required,   176-178
  description of,   4, *see also* Labor
  distribution of, across activities,
    178-180
  expenditures,   180, 184
  management system element,   5
  profile,   176, 188
  summary curve,   176, 188
  units of measure to determine,   180
Responsibility, assigning,   40-43

S
Sample sheet,   71
Schedule. *See also* Scheduling; Tasks
  adjusting, to improve the resource
    expenditures,   180, 184
  documentation,   170
  disorganized, problems with,   124
  levels of detail,   124-126
  organizing the,   123-139
  primary reasons for designing and
    organizing,   123-124
  reporting and updating the,   171-172
  summary, for upper level management,
    171
  target, establishing,   153
  update, monitoring progress and,
    144-152, *see also* Schedule Update
Schedule records
  initial development phase,   159-167
  maintaining,   159-172
Schedule reports,   243
Scheduler, professional,   49
Scheduling. *See also* CPM; Schedule
  basic steps of,   47
  construction. *See* construction schedule
  hand, system,   127
  introduction to,   47-49
  keeping good records,   42, 70-71
  maintaining up-to-date,   8

over-complex,   49
as the primary tool of the project
  manager,   ix
process, breaking down into sub-parts.
  See sub-tasks
procurement activities,   119-120
the project. *See* Project, scheduling
step-by-step process of setting up,   47
second phase of activities,   48
unrealistic,   49
working files,   41
Scheduling update
  individual activities,   145-147
  monitoring progress and,   144-152
  problems with the various methods,
    147
  steps in,   145
Shop drawing and procurement. *See*
  Procurement Site
  job. *See* Job site
  types of,   35
Subcontractor
  arrangements,   35
  priorities and obligations,   108
  process for dealing with,   36-37
  scheduling,   54-55
  scheduling, as a problem for contractor,
    8
Submittal data. *See also* Procurement
  control log,   120
  coordinating, with schedule,   111-119
  definition of,   108
  following up on the,   120
  keeping a log of, approvals,   111
  management of, and procurement,
    107-120
  procedures for,   110
  reporting,   120
  status lists,   120
Submittal list, making a,   110
Sub-tasks. *See also* Activities
  breaking the job down into,   53
  description of,   48
Suppliers, contacting,   37-38
Systematic management,   3-4

T
Tasks
  information gathering,   31
  major project management,   4
  of the project manager,   xv, 6
  required to provide schedule
    information,   126
  sub. *See* Sub-tasks
Time. *See also* Project time
  checking progress of,   xv
  cost trade-offs,   195-203
  management system element,   5
Time cards,   219
"Time now" date,   145
Time, project. *See* Project time
Tracking system
  labor and equipment costs,   218
  monitoring construction activities
    through,   119

purpose of, 109
Transmittal letters, keeping good records, 41
"Trial balloons," 71

U

Unions and labor suppliers, 39

W

Work
  direct measurement of, 218
  establishing the sequence of, 57-58
  office, construction managers and, 107
  scope of, 37
  tips for establishing sequences, 63-65
Workdays
  definitions of, 70
  conversion of, to calendar dates, 101-103
Working files, 41

Z

Zero duration, 92